It's About TIME

Creative Activities About Time

by Robynne Eagan and
Tracey Ann Schofield

illustrated by Wendy Grieb

Teaching & Learning Company

1204 Buchanan St., P.O. Box 10
Carthage, IL 62321-0010

This book belongs to

_____

Cover by Wendy Grieb

Back cover photo of Tracey Ann Schofield by Glamour Shots ®

Copyright © 1997, Teaching & Learning Company

ISBN No. 1-57310-073-0

Printing No. 987654321

Teaching & Learning Company
1204 Buchanan St., P.O. Box 10
Carthage, IL 62321-0010

TLC10073 Copyright © Teaching & Learning Company, Carthage, IL 62321-0010

Dedication

To all of the children who help us to understand that time is a relative thing. R.E.

This book is dedicated to my dearest friend, Judy, whose time on Earth ran out too soon; my three extraordinary children, Matthew, Patrick and Stephanie, who have convinced me that there is no time like the present and no present like time; my wonderful parents, Barb and Paul, with whom I have shared so many good times; and my long-suffering husband, Jonathan, for whom I seem to have so little time! T.A.S.

Acknowledgements

Thank you to the ancient sky-watchers who first recognized the patterns of the universe and used these to help keep time, to the first inventors and scientists who helped to bring simple, consistent timekeepers to our walls and our wrists and to the educators and parents who taught us about teaching kids. Thank you also to the people who put so many hours into writing the papers and books and compiling museum exhibits that lead us all to a better understanding of time. And to Mark for his technical assistance, Michaela for sharing her teaching and writing expertise, Denise for her time management skills and Charles for his scientific research. Finally, thanks to Thomas Eagan (M.Sc.Physics) for verifying facts and sharing his knowledge of space-time and physics.

Table of Contents

Dear Teacher or Parent,

The passage and measuring of time has been a source of intelligent inquiry and provocative investigation since humankind first left its tracks on the Earth. Now, there's a resource tool to help children unlock the mysteries of time in the comfort of their own home or classroom.

It's About TIME is a child's guide to the discovery and mastery of time. The resource is designed to help parents and educators use time to teach learners in grades 1-4 about life on our planet. It's About TIME opens doors to active learning opportunities that encourage kids to formulate questions, investigate their environment and make exciting discoveries. Anyone can use this resource to tap into the natural inquisitiveness of the young mind and satisfy a child's need to bring order to a confusing world.

Ten fun and fact-filled chapters feature guided discussions, fast facts, quick queries and hands-on investigative learning activities. Material is presented in an easy-to-use format with at-a-glance details, concise background information and clear activity directions. This open-ended topic is easily integrated into the math, science and history curriculums with skill, concept and fact reinforcement that can be extended into other areas.

From timely experiments to classical clock watching, children will learn to keep time, measure time, tell time and understand its effect on everything in the universe. They will measure and tell time by the sun; the clock and the calendar; find out why there are 24 hours in a day and seven days in a week; travel to the past, present and future and learn about the body clock that ticks within each of us.

From the ancient civilizations who first tried to understand, arrange and measure time to modern society where we organize our lives around the clock—time is a fascinating topic that affects every one of us. Children experience time every minute and second of their lives—it only makes sense that they should become timewise.

Sincerely,

Robynne Eagan and Tracey Ann Schofield

Symbols and At-a-Glance Information

Throughout this book you will find the following at-a-glance information to provide you with important details regarding the activity or lesson.

Grade Levels
Activities will pertain to grade levels from 1-4.

Curriculum Links
The information and activities in this book can be extended and incorporated into several curriculum areas to facilitate development of various skills. Curriculum areas specifically targeted through these activities will be indicated as such:

Math	Language	Design and Technology
Science	Geography	Health
History	Physical Education	Art

Skills
Activity headers will list skills that will be developed or enhanced through the exercise.

Time
An approximation of the time needed to complete this activity will be indicated by:

10 min   30 min   60 min

Caution
Particular tasks may require adult supervision.

Fast Facts
Fast facts can be found in **bold** text throughout the book. These facts provide interesting trivia to bring a topic to life or to further elaborate on relevant details.

Quick Queries
Quick queries can be found identified as such and *italicized* throughout the book. The queries are devised to stimulate thought, reinforce skills and enhance children's understanding of concepts.

A Word About Hands-On Activities
Simple procedures and everyday materials will make it easy to incorporate hands-on investigations to enhance the learning process. Set rules regarding conduct and the safe handling of equipment and materials before engaging in any activity.

Dear Parents/Guardians,

Your child will be learning about time; the passage of time and the measurement of time. This cross-curriculum unit will introduce children to the mathematical concepts involved with time and help with the development of skills in math, science, language, history and geography skills. Your child will develop a better sense of time and time passage; build a time-related vocabulary; learn about the past, present and future and, of course, learn how to tell digital and analog time.

To aid in our look at time, we need the following items for our classroom:
- cotton balls
- paper cups
- funnels (labeled so these can be returned)
- homemade funnels (remove the bottom from rounded plastic bottles)
- sand timers (labeled so these can be returned)
- stopwatches
- electric timers
- wind-up timers
- plastic soda bottles
- old clocks, watches or other recovered time-keeping devices
- pictures of time-keeping devices
- books about time or clocks

If you have any interesting "timepieces" you would like to share with the class or if you have a suggestion or idea to share, I would be pleased to hear from you.

Thank you for your assistance. Your contributions will help make this unit fun and exciting.

Sincerely,

Chapter 1
What Is Time?

You've probably heard the question "What time is *it*?" often enough, but have you ever been asked "What is time?" That is a much more difficult concept for children to comprehend. Children can't touch time, look at pictures of it or watch it move from place to place as it passes.

Throughout the primary school years children seem to develop a natural sense of time just from being alive. In grades 1-4 children begin to recognize the important role time plays in our society. They are aware that there is a system to measure hours and minutes and days and years. They are often familiar with the time terms *year, month, week, day, hour* . . . right down to the millisecond. Generally by about the age of nine a child has developed a good awareness of and sense of time.

So, what is time? We know that time is in constant motion, flowing like a river from the past to the present in an established forward direction at a rate of 24 hours per day.

We know a lot about time, but it could be said that there is no true answer to the question "What is time?" Throughout history and around the world cultures in different eras have viewed time in diverse ways. Although we cannot precisely define *time*, we can help children to develop a much better understanding of time by discussing it, exploring and experiencing it, by thinking about it and finding out what is known about time. This book will help children come to understand the measurement of time from centuries to seconds and help them to comprehend the more elusive qualities of the passage of time.

Time Terms

Interval: The length of time between the start and finish of something.
Epoch: A specific location in time where an event will or has occurred.

Fast Fact: Chronos, the ancient god of time, was often referred to as the "soul" of the universe.

Fast Fact: The great thinkers of ancient times believed that time was the ultimate judge who would discover and avenge injustices.

Guided Discussion

The best way to start talking about time is to . . . well, talk about it! We experience time every living moment and each of us has our own unique understanding of it. Find out how much the children already know and understand about time and take it from there.

Use the following questions as guidelines to find out what children think and know about time. Children of all grade levels can brainstorm, discuss, think about and answer the following questions in age-appropriate terms. You may be surprised to discover the extent of abstract thought children are capable of.

- *What do you know about time? Is it important in your life?*
- *Can we measure time?*
- *What do you use in your life to help you to keep track of and measure time?*
- *What do we use to keep track of the days? The weeks? The months? The years?*
- *What do we use to measure hours and minutes and seconds? How long is an hour? How long is a minute? How short is a second?*
- *Have people always measured time the way we do today? What do you think was the first thing people used to help them tell time?*
- *Does it sometimes feel like time has slowed down? Is time really going slowly? Why do you think it feels that way?*
- *Does time sometimes pass very quickly? What are you doing when it feels like time is passing quickly? Is time really passing more quickly? Why do you think it feels that way?*
- *How much time has passed since you were born?*
- *What is a long time? What is a short time?*
- *Can we move forward through time? Can you control the speed or direction of your trip through time? Can we speed it up or slow it down if we want to? Can you stop moving through time if you want to? Can we go backwards through time? Can we travel to other times?*
- *What do you want to know about time?*

The River of Time

Time is often described as "flowing" and can be visualized as a stream that never ends. Time does flow on and on in one direction and carries us along with the current. We can conceptualize the passage of time throughout our lifetime as a journey along the river of time. From birth to death we are swept up by the current of time and moved along at a steady rate through time. We can't go back in time to where we have been–we just move forward with the flow of time to the next bend in the river–around which we cannot see. We know what happened upstream where we already were, but we don't know what is going to happen downstream–where we're headed.

Materials

reproducible on page 12
pencil for each child

What to Do

1 Hand out a pencil and "The River of Time" reproducible (page 12) to each participant.

2 Direct children to listen carefully to the following instructions.

Pay attention to the following instructions. I will repeat each instruction twice so please listen carefully.

1. Put your pencil on the empty circle in the upper right-hand corner of your page.
2. Print the number one in this circle.
3. Move your pencil to the next empty circle traveling downward and to the left. Print the number two in that circle.
4. Write numbers in the circles in this way–moving down and across your page until you have numbered the circles all the way to number 10. Fill the last circles with the numbers 25 and 50.
5. These numbers represent years in your life. Find the spot on the river that represents your current age. If you are 8, draw a boat near the circle numbered 8–if you are $8^1/2$, draw a boat between 8 and 9.
6. Draw yourself in the boat as you are now.
7. From your position in the boat you can see upstream with your memories to places and events in your past. Try to recall what you saw and did and felt as you flowed down the stream of life.
8. You can't see ahead to places you have not yet passed, but you try to imagine what might lie downstream for you. Think about what you might be doing or hope to be doing at the ages beyond where your boat now lies. Fill the river with predictions and hopes and dreams for these future years. Then color your picture if you wish.

The River of Time

ONE WAY

43

Time Is a One-Way Trip

Unlike space, which allows us to move in any direction–up, down, forwards, backwards, sideways–we can only move one way through time: forward. This direction is called the fixed arrow of time.
In the pictures below, the sequence (or timing) of events has been mixed up. Can you put time right?

Below are pictures of five different sequences. The order of the pictures has been mixed up. Cut out the pictures for each sequence. On another piece of paper, paste the pictures in the proper order.

Time Keeps Rolling On

Materials

roll of adding machine paper
clock
marker

What to Do

1 Lay the roll of paper down flat. Place the clock at the beginning of the roll of paper and unroll it away to left.

2 Roll the paper out 1" (2.5 cm) at a time and mark one hour for each inch (centimeter).

3 This paper is like a river of time that flows by-hour by hour.

4 Continue marking until you have one full day and are beginning the next day on the paper. Each mark on the paper corresponds to a different place along the river of time. From any point on the paper you can look back and recall when you wrote the previous hours, but you cannot recall writing the hours to come on the paper that are still in the roll. It is not possible to recall events from the future.

Patterns in Life

Children develop a natural sense of time by observing repetitive patterns in their life experiences. They become aware of patterns that occur on annual, seasonal, monthly, weekly and daily intervals.

Young children can develop a better sense of time by becoming more aware of the natural cycles and human routines that define our lives. You can draw their attention to the patterns and sequences of events that comprise their lives.

Time Passages

When we know what schedule or clock an event is timed to, we can note the passage of time *relatively* by thinking of other things that take the same amount of time.

For example, if we know that something is going to take a month, we can think of that time in a number of ways that might make it more meaningful. One month is the time it takes for: the moon to get full again (assuming today is a full moon!); Mom to go grocery shopping twice; four swimming lessons; 20 school days; 12 baths; the list is infinite.

Use things that are meaningful to you to think of three different ways to describe:

• one hour • one day • one week • one month • one year • one lifetime

We Are All Time Travelers; The Past, Present and Future

We use language to help us understand every concept, so time is no different. There is a large vocabulary–almost a language in itself–that has arisen to help us understand time. We tend to discuss time by using the terms *past*, *present* and *future*. An understanding of these terms leads to a deeper comprehension of the nature of time.

From the time we are born we travel forward in time whether we want to or not. We all once lived in the past, traveled to the present and will one day visit the future . . . only then it will be the present! It could be said that we are all time travelers with no control over how fast or which direction we travel!

The Past

We call time that has passed "the past." Time moves forward; a day moves from morning to night and never in the opposite direction. Spaces and places can be revisited but time cannot. Eating, sleeping, working and playing occur in moments snatched in time and if you miss an opportunity you can't have another chance. When a moment has passed, there's no going back.

Quick Query: *How long ago did these events happen? The formation of the Earth, your parents births, your birth, your first day of school.*

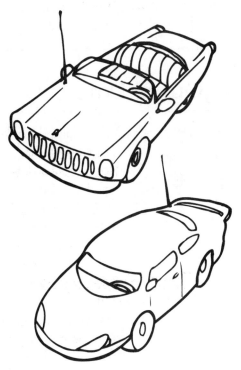

How Far Back Can You Remember?

Take a look through your memories. Can you remember something from the past? Do you remember when you were born? Do you remember when school started? Can you remember when the Earth was formed? Memories help us to remember times that have past. All of these events happened in the past but some happened long before others!

How Old Is Old?

With the passage of time, things and people begin to get older, or age. Things that are young or new change and become old as time passes. Some things change a lot and some change very little.

1. Look at buildings, books, toys, relatives, cars and so on. Talk about how these things change as they become older.
2. Fold a strip of paper into four quarters. Draw a sequential diagram to show how something has changed over time.

Make a Time Line

A time line is a linear picture that represents the chronological passage of time noting important events that occurred along the way. Have children make time lines to represent their life or your school year together–choose the intervals you will mark.

Quick Query: *The Greek word **chronos** has come to mean "time" and is used in many time-related words. Can you think of any such words?*

15

Faces from the Past

Can we look into the past? Well, sort of. We can look at photographs, videos, recordings and our own memories, but we can't go back into the past to watch events that have already happened. What has passed is in the past!

Materials

chalkboard display area and chalk
baby and current photograph of
    each participant
Blutak™ or tape

Get Started

Have each child bring in a current and baby photo of himself. Place the pictures on the chalkboard using Blutak™ or tape. Place the current pictures on one side of the board and the baby pictures on the other. Number all the pictures on one side and then the other.

Object

To match the baby picture with the current picture.

How to Play

1 Have participants draw chalk lines to connect the pictures they think match or record their matches on paper using the numbers assigned to each picture.

2 The exercise can be just for fun or players can compete to see who matches the most pairs correctly.

Try This

• Follow up by attaching yarn from each baby picture to the corresponding child picture.

Out-of-Date

When things have been a part of the past and are no longer "trendy," we say they are out-of-date. Host an out-of-date day where children wear clothing, shoes and other fashion accessories from the past. Children can contribute to a display of out-of-date coins, stamps, magazines and other memorabilia. Share some out-of-date music and out-of-date treats. But be careful–some things come back in style after a time and your out-of-date party might start a trend all over again! Chances are this exercise will facilitate communication between parents, grandparents and children and help children to learn about the past.

Long Ago

Encourage children to think about the past by copying and completing the sentence: Long Ago . . . They might discover that long ago to one person may not be long ago to someone else.

There's No Time Like the Present

What is the present? It is right now! If we want to be very particular it becomes quite an elusive moment that quickly slips into the past. Sometimes we refer to our present time or era as "the present."

Yesterday is a memory, tomorrow is a dream,
today is the reality . . . make the most of it.

Live for the Moment

We can't go back to yesterday, and we can't travel into the future, but we can be in the present–in the moment we are living right now. Do you think about the past or the future very often, or do you live for the moment? Most children are pretty good at living for the moment–living in the present that is. It is important for all of us to appreciate and savor the moment we are living in–the "now."

Quick Query: Ask children to stop for about 15 seconds and think about the moment. What do they see, hear, smell, feel and taste?

Capture the Moment Bulletin Board

Create a Capture the Moment bulletin board featuring children's haiku poetry and prized photographs.

Introduce children to Japanese haiku poetry. Help them to feel the immediacy, of capturing a moment in very few words. Have children capture the moment in their own 17-syllable haiku poem. Frame the poems in construction paper cut-outs of items that represent the poems.

Discuss the fine art of photography. It captures a moment on film. Provide a camera for classroom usage that allows children the opportunity to capture moments for themselves. Display these photographed moments along with the written moments.

The Present Era

Discuss the present era. What things do children think represent their current era? What things do they think will go out of style? What do they think will last into their children's time?

Our Era Collage

What do kids think is important in their era? Have children search through magazines and newspapers to cut out items that represent their era. Paste these cut-outs to mural paper. Cover the entire area. Paste the year in the center of the collage.

The Future

Children have been in the past and are living in the present but can't travel to the future. What is the future? Any time beyond the time we are currently in is called the future. To be very specific, the future can be one second ahead of where we are (or by now were!) or looking at the future in the context of a larger time frame it implies anytime yet to come – usually days, weeks, years, centuries or more.

Growing into the Future

1-4 • Science/Math • Time and Linear Measurement/Information Recording/Experience with Symbols/Observation • 60 min

Fast-growing bean seeds can help us to understand the passage of time. Living things grow bigger and older and change in many ways with the passing of time. Growth helps us to grasp concepts of the past, present and future.

Materials

soil
pots
green bean seeds
small shovels or spoons
calendar
marker
2" x 24" (5 x 61 cm) strip of sturdy paper
masking tape

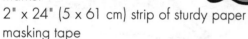

What to Do

1 Children plant a bean seed in a paper cup, using soil and a small "shovel" or spoon.

2 Place cups in sunlight. Water the seeds thoroughly. Mark a small shovel on your calendar to represent Planting Day.

3 Continue light waterings as needed and observe the seeds each day.

4 Choose one cup to be used for the calendar record. Place the cup on the chalkboard ledge. Fasten the measuring strip on the board behind the plant. Position the start of the tape at the top of the soil.

5 Draw planter cups showing no growth for each day of the calendar when you cannot see any sign of the plant. Draw cups to represent small and then larger sprout growth.

6 Use the measuring strip to measure the growth of the plant. Record the measurement inside the "cup" on the calendar each day and record the date beside the measurement on the strip.

7 Keep an ongoing record of the growth and changes of the plant.

8 Recall the day in the past when you planted the seed. What day did the seed first sprout? On what day was it a particular height? How does the seed look today, in the present? What do children think will happen to the plant in the future?

In the Future

Is it possible for you to know what will happen beyond the present, in your future? You can make some predictions, but you can't know the future before you get there and then it becomes the present and in no time at all it is in your past! Make some predictions about *your* future!

Tomorrow I will _____

_____

Next year I will _____

_____

I will need a haircut in _____ days.

In 20 years I will be _____ years of age.

I think that I will be _____

I will have _____

_____

I will drive _____

I will live _____

The biggest change in the world will be _____

_____

In 20 years my town will look like this:

Time Travel: It Could Happen to You

Have you ever wanted to travel in time? Maybe go back to the land of the dinosaurs or zoom ahead to the year 2001? Believe it or not, some scientists believe that time travel, at least in theory, is possible. Certainly, there is nothing in the laws of physics that would prevent it.

For now, time travel is the stuff of fantasy and fiction. But someday we might have the cosmic technology required to build a fully functional time machine. This would include the tools that would allow us to:

- gather together at least 10 stars with the mass of our sun and squeeze until they collapse and create a black hole in space
- plunge in one end of a "wormhole," or "space-time tunnel," and burst out the other
- twist the wormhole by dangling a planet at the two "mouths"

Time-E-Scapes

A Blast from the Past

3-4 • History/Language/Science • Information Recall/Problem Solving/Creative Thinking • 10 min

So far, your mind is the most powerful time machine we know. It's easy to use–just give it a try!

1 Using mind travel, go back in time. It could be an hour, a day, a month, a year.

2 Change one event. Maybe . . . instead of getting a puppy when you were six you didn't.

3 Then move forward in time again and see how that one change in the past would change everything that happened afterward.

4 Write three good things that happened because of that one event that would not have happened if you erased that event from your history.

5 Write three bad things that happened because of that one event that would not have happened if you erased that event from your history.

Day Trip to the Future
1. Now, travel into the future with your mind machine.
2. Use your imagination to think of three future events that might happen as a result of that one event in the past.
3. Think of three things that might not happen because of the event.

Time for a Change
Imagine what you would do if you could go back in time and change things, even just a little. You would be the most powerful and influential person on Earth.

1. Imagine what would happen if you could erase the existence of one person–Adolph Hitler, for example–from the history books. What effect would that have on world history?
2. What if Henry Ford had not invented the automobile?
3. What if whatever caused the extinction of the dinosaurs had been prevented?

The Beginning of Time

Have you ever heard the phrase "from the beginning of time"? Did you ever wonder what it means? It implies that there was a beginning time and that is a difficult concept to think about. We can say there are several perceived beginnings to time.

Quick Query: **When do you think that time began?** *Invite children to think about and discuss this question before presenting the various possibilities.*

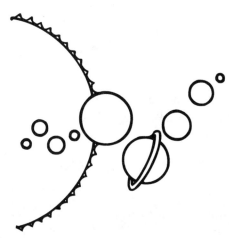

The Beginning of the Universe

Astronomers consider the beginning of time as the beginning of our universe—that is generally believed to be 15 billion years ago. Scientific evidence indicates that there was a definite beginning to the Universe and it has been expanding and getting older ever since.

The Beginnings of Earth

Scientists think the Earth, like everything else in the solar system, was formed from a huge cloud of materials floating in space about four and a half billion years ago. It's very, very difficult to imagine a time that long ago! If you consider the formation of our planet to be the beginning of time—you could say time began about four and a half billion years ago.

Time Began When . . .

Some people believe that time, as we know it, began with a very important event. The events vary between cultures but are often based on the birth of a great leader or savior. Civilizations began measuring time from the date of particular events. In our system of marking the years, the year 1 notes the birth of Christ.

Fast Fact: We live in the 20th century and will soon enter a new century and a new millennium. Because our calendar time began in the year 1 the 21st century really begins on January 1 in the year 2001. In that year we can say that time began 2,000 years ago. Many people will be celebrating the year 2000 even though "our time" will only be 1999 years old on that date. When will you celebrate?

Quick Query: *How old will you be in the year 2000 A.D.?*

If Time Has a Beginning, Does It Have an End?

Time seems to have a definite beginning and end for human life but maybe not for the universe. The universe has been expanding ever since it began 15 billion years ago, and time has been moving forward ever since. Scientists expect the universe to continue getting older for a long, long time.

Time Is Like a . . .

1-4 • Language • Writing/Creative Thinking/Information Recall • 30 min

1 Have children complete the sentence "Time is like a . . ." using only one word. Record the responses on chart paper.

2 Have students write a short essay on this topic.

Space and Time

In 1905 a clerk in a Swiss patent office published a paper that changed the way scientists thought about time and space. That clerk was the German physicist, Albert Einstein – you may have heard of him. The paper was called The Special Theory of Relativity and it was followed by another paper in 1915 called The General Theory of Relativity.

Before Einstein's papers, space and time were thought about as different entities. Einstein's theories brought the concepts of space and time together.

Space is what surrounds us in all directions – left and right, forward and backward and up and down. Space is a three-dimensional concept that we can see – from here to there, and measure as feet, inches, meters, miles, whatever.

Time is more difficult to define; we can't see both ends at the same time and we can't control our movement through time. We always travel forwards (not backwards, sideways or up and down) through time, passing the minutes, hours and days in a chronological fashion.

Space and Time Together

Einstein made the world think about time in a new way by combining the concepts of space and time. Bringing these two concepts together takes some mind power!

People used to think that time was a constant – that it passed at the same rate for everyone, everywhere. Einstein's general theory of relativity predicts that in fact, time would pass more slowly if you could travel at very, very high speeds (near the speed of light). Einstein's theory of special relativity predicts that time would also pass more slowly if you could go where there are very strong fields of gravity. In both cases time changes – it slows down. Experiments have proved these basic assumptions to be true.

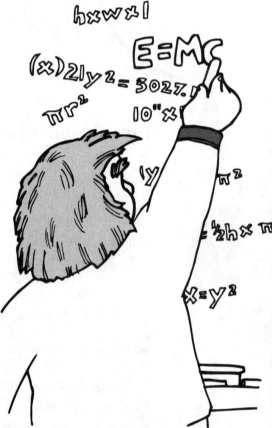

Fast Fact: Scientists believe that time may pass slower and slower with an increase of gravity and may even come to a complete stop in the great gravitational field of a black hole deep in space.

Quick Query: According to Einstein's theory, would you age faster or slower if you lived on a very, very high mountain with low gravity? What would happen if you were able to travel on a vessel faster than anything in our era can be made to travel?

The Fourth Dimension

In the 20th century, after Einstein presented his theories, scientists began talking about a four-dimensional space called space-time. It isn't possible for us to see time spread out the way we see space spread out – but there are mathematical equations to explain events that occur in space-time. Think about an event like a ball game to be played at a particular point on Earth. You know the location of the stadium, but unless you know the time – you're sure to miss it!

Measuring Space Distances Using Time and Light

Quick Query: Try to imagine the vast distance around our Earth, from Earth to the moon, from Earth to the sun, from the sun to Pluto and then beyond through our galaxy-and then on to other galaxies! How would you describe these distances?

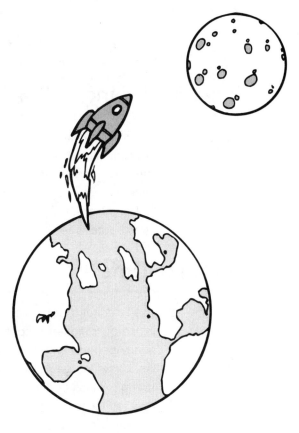

Scientists use the speed of light as a unit for measuring space and time. No matter where light travels, in water, air or through space, it always travels at 186,000 miles per second-it is called the speed of light, and because it doesn't change we say that it is a "constant." The distances in space are so vast that it is almost impossible for us to imagine. Scientists used the speed of light to help them measure these great distances, which aren't measured in feet or miles, but in measurement called light-years.

The time that light spends as it moves through space becomes a measurement of distance-bringing the concepts of space and time together in the unit of the light-year.

Fast Fact: A light-year is the distance that a ray of light can travel in one year in a vacuum-5.88 trillion miles.

Looking into the Past

It takes time for light and sound to reach you so everything you see or hear actually happened in the past. If a friend shouts to you from far away and shines a flashlight, you hear and see things from the very recent past.

When you see something light has traveled to your eye from another source. Can you see things in the dark? Because it takes time for light to travel and reach your eye-in a sense you see things from the past. Some things you see are from long, long ago-like when you look at a star. It takes light-years (which is a very, very long time) for the light of a star to travel through space and reach your eye. You may, in fact be viewing a star as it looked millions of years ago! It is even possible to see a star that no longer exists!

Fast Fact: Light from our sun takes eight minutes to reach Earth. The sun you see is the sun as it existed eight minutes ago. (Of course, you should never look directly at the sun!)

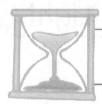

Chapter 2
Lifetimes

Looking at life spans of living things can help children gain some perspective of time. The natural world is a continuous cycle of new life, growth and decay. All living things have a life span.

The Beginning: "The Baby Imperative"

For all living things, nothing is more important than producing the next generation, either as seeds or babies. There is a right time to start a baby, and this follows a regular cycle over a number of years. The cycle and its length differ among species. Tiny microscopic organisms called bacteria can produce millions of offspring during their lifetime; humans can only produce a few. The number of young that one individual can produce is dependent upon factors including: gestation length, nurturing time, growth rate, age of maturation and lifetime.

Tummy Time

The time that a baby mammal spends developing inside of its mother is called the gestation or pregnancy period. In general, the larger the mammal, the longer the gestation and the fewer the offspring. A mother rhinoceros, for example, is pregnant for 15 months before giving birth to a single baby, while the mother rabbit, who has up to eight young in a litter, is only pregnant for one month.

Fast Fact: The Asiatic elephant has the longest gestation period of any mammal. On average these mothers are pregnant for 20 months! The shortest gestation period of only 12-13 days, is that of the Virginia opossum of North America and the water opossum of South America.

Some very tiny living things, such as single-celled bacteria, multiply by splitting in two. In certain bacteria, the time from one split to the next is only 20 minutes. These microscopic creatures grow and mature so quickly, they can increase their numbers incredibly fast. A single bacterium can produce a quarter of a million offspring in 10 hours!

Fast Fact: If a single cabbage aphid was supplied with a safe environment and unlimited food supplies, it could produce 906 million tons of descendants in one year!

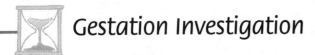

Gestation Investigation

Take a trip to the library and research the gestation time of three different animals. Try to find animals that are small, medium and large.

Life Spans

Every living organism seems to have a built-in time-keeping device that determines its general life span. Small creatures tend to live their lives more rapidly and die sooner. Large animals lead a more leisurely life-style and tend to live longer.

Every living thing has its own life span and growth rate. All living things grow change over time – some grow and change very quickly, while others take much longer. The time that it takes different creatures to reach maturity (the age at which they can produce babies of their own) also varies. It can be as little as 20 minutes or as long as 12 years.

Small animals grow fast and reach maturity quickly, but they do not usually live long. Their clocks tick quickly, and they produce as many babies as they can as fast as they can. Insects and small mammals such as mice and rats follow this kind of life plan. Even though they do not get very big or live very long, their way of life is extremely effective.

Bigger animals produce fewer young, but they live longer. For this kind of life pattern, size is a definite survival advantage. It is easy to frighten, ignore or eat smaller animals if you are large, and big animals with their big fat reserves are better at surviving the cold than smaller ones. Because bigger animals live longer, they have more chances to reproduce. Of course, there is a price to pay for this bulk. It takes a very large animal such as an elephant, a long time to grow large and bear young. The survival of each baby is so important, large animals must look after their little ones carefully and for longer periods of time than their smaller counterparts. This means that large animals – such as bears, gorillas and the largest birds and reptiles – increase their numbers relatively slowly.

Animals that take more than 10 years to grow up include human beings, elephants, Nile crocodiles and the 17-year cicada an insect that takes – you guessed it – 17 years to reach maturity!

Observing the changes in the things around us is one way of noting the passage of time. We often use the growth of things around us as indicators of the passage of time.

Quick Query: *What do you think is the typical human life span?*

The Human Life Span

Being human ourselves, a look at this life form is probably of the most interest to us! What is the usual life span of a human? Have children think about people in their lives. Who is young? Who is old? Who is the youngest and who is the oldest person they know?

Growing at Your Own Pace

Every living thing grows at its own speed. Some things, like sunflowers and bean plants grow very quickly–in days, in fact. Some things take a little longer to grow but still grow quite quickly, like baby ducklings that take eight weeks to become full grown. Some things grow very slowly, like humans, which grow in years or oak trees, which take decades and even centuries to grow.

Growth Chart

K-4 • Science/Health/Math • Comparison/Measurement/Concept Reinforcement • 10 min

What to Do

1 Draw three columns and label them *Slow*, *Medium* and *Fast*.

2 Have children write names of, draw or paste pictures of things considered to grow at the pace indicated by the column header.

3 Have children bring in a photo of themselves as a baby. How long does it take for a human being to grow before being born? How long does it take to grow from a baby to a child? From a child to an adult? From a young adult to an old adult? Under which column do children think human beings would fall?

A Year

We measure the growth or aging of people in intervals of one calendar year–365 days. That year between one birthday and the next may seem like a long time!

Fast Fact: The average American woman will live to be 79.
The average American man will live to be 72.

Age Line-Up Activity

K-4 • Math/Physical Education/Health • Problem Solving/Group Interaction • 10 min

Materials

group of six or more
musical tape or radio

What to Do

1 Have students form a line that starts with the youngest child and ends with the eldest. The children who were born first have lived longer and are older than children who were born later.

Try This

• Have children move to music and then regroup in this line as quickly as possible.

Name _____

My How You've Grown!

How Old Are You? _____

Sam   Rosa   Masud

Look at these three children.

How old do you think they are?   Rosa _____   Masud _____   Sam _____

Who do you think is the oldest?   _____

Who do you think is the youngest?   _____

Is Rosa or Masud older than Sam?   _____

Are you older or younger than Masud?   _____

To see how much you have grown and changed since you were a baby, compare a recent photograph of yourself to one taken when you were an infant.

Do the two photographs look very different? _____

How have you changed? _____

_____

_____

When Will I Grow Up?

Minutes

If you were a bacterium, you would be fully grown and ready to divide in two in a matter of minutes.

Days

If you were a bean sprout, you would grow from a seed to a tiny plant with roots and leaves in a matter of days. See for yourself:

 ## Sprout Growth

K-4 • Observation/Recording/Calendar/Skills • 30 min

Materials

bean seeds
jar (with lid) of water
sink for rinsing

What to Do

1 Soak the bean seeds in the jar of water and cover tightly with the lid.

2 For the next week or so, take the lid off the jar and gently rinse the seeds every day, making sure to change their water each time.

3 Observe! After only a few days, they should begin to sprout. In about six days, they should be little plants.

4 Chart the plant progress on your daily calendar.

 ## Weeks

If you were a fruit fly, you would be fully grown two weeks after you were born.

Months

If you were a kitten or a pig, you would be ready to start a family of your own after just six months.

Years

If you were a human begin . . . Wait a second! You *are* a human being, and you will not be considered to be a "mature adult" until you are at *least* 18 or 19 years old (although your body will be physically prepared to produce babies by the time you are about 12).

Quick Query: *Can you think of at least one creature that takes minutes, months, days, years, weeks to grow?*

Name _____

How Times Have Changed!

When you were very tiny, your age was measured in days. After 14 days or so it was measured in weeks. After about 12 weeks, it was measured in months and now it is measured in years.

One day your age might be measured in terms of decades (groups of 10 years). If we talk about some-one who is 20 something, for example, we say that he is in his twenties. If we wish to be a little more specific without giving a number, we can say that she is in her early, mid or late twenties (early: 20, 21, 22, 23; mid: 24, 25, 26; late: 27, 28, 29).

Your Age

Are you one decade old?_____

Are you in your "teens"? _____

Are you one century old? _____

Give these people an age.

Aunt Juanita is 26. She's in her?_____.

Uncle Dante is 38. He's in his?_____.

Great Aunt Jill is 41. She's in her?_____.

Great Uncle Andrew is in his mid-fifties. What ages could he be?_____

Grandma Jessie is in her late sixties. What age might she be?_____

Grandpa Mario is in his early seventies. What age could he be?_____

Draw a line to match the age to the name.

Newborn Nancy   late seventies

Teething Tony   5 years old

Skipping Sally   11 months old

Teen Tenisha   mid-thirties

Mom Maria   6 days old

Grandpa Gary   18 years old

Challenge

How many whole decades are there between Grandpa Gary and Newborn Nancy?_____

Family Tree

Children will enjoy learning about their ancestors as they explore the past.

Materials

construction paper
markers
paste
scissors
photographs, photocopied
   photos or drawings of
   family members
Family Tree Information
   Sheet (page 31)

What to Do

1 Review the Family Tree Information Sheet with the children.

2 Explain that the tree is a way to put the child's family together so that it is easy to understand. Have each student make their own unique family tree transferring the information provided by their Family Tree Information sheet to a "tree." Have children draw a tree trunk. They write their name on the trunk. The top of the trunk will contain names of siblings. The primary branches will contain the names of their parents and their siblings. Cousins can be included as apples hanging on the tree. If the child has information on two parents, the tree can be divided in half in some way; dark green/light green, high branches/drooping branches. Branches that rise above these will include the names of grandparents and great-grandparents.

Get Started

1 Ask children to think about their family or guardians, people in the present to whom they are connected to.

2 Now ask children to travel with you back to the past and think about their family. Who was their mother/father/guardian's mother or father? Did they have brothers or sisters? And who were their grandparents parents?

3 Send home the Family Tree information Sheet on page 31. Allow two weeks for children and families to talk, dig out old photos and share family stories.

Try This

• Children can decorate the tree and surrounding area to represent their family.
• Advanced students can include further details regarding grandparents and their siblings.

Fast Fact: A generation is the time frame of about 30 years that it takes for children to grow up and step into the shoes of their parents. This also refers to a group of people in the same age group who share similar experiences and opinions.

Family Tree information Sheet

Dear Parent/Guardian,

We are going into the past to learn more about our families and to better understand time. Please help your child to complete the following chart by _____. You may choose the mother, father or guardian's branches of the family tree to explore for this project.

Please share a family story and, if possible, send a memento of your family—sealed in a plastic bag and labeled. Although we will take every precaution with these items, please do not send valued or irreplaceable treasures.

Thank you for time and interest in this project. I hope you enjoy sharing your family history!

1. Child's Name _____ Birth Date _____

2. Siblings _____

    _____

3. (Parent or Guardian _____

4. _____'s Siblings and Children _____

    _____

5. _____'s Father _____

6. _____'s Mother _____

    _____

Second "branch" of the family tree (optional)

7. (Parent or Guardian _____

8. _____'s Siblings and Children _____

    _____

9. _____'s Father _____

10. _____'s Mother _____

The Clock of Evolution

Scientists have a good idea about when different evolutionary events took place. The major biological milestones are listed on the clock below. This clock of evolution can help to put the existence of our species into perspective. Compared to the Earth, humankind has been around for a very short time indeed!

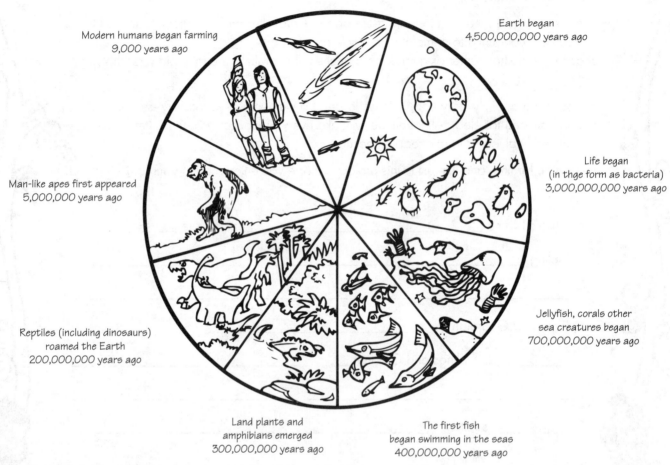

Universe created
15,000,000,000 years ago

Earth began
4,500,000,000 years ago

Modern humans began farming
9,000 years ago

Life began
(in thge form as bacteria)
3,000,000,000 years ago

Man-like apes first appeared
5,000,000 years ago

Jellyfish, corals other
sea creatures began
700,000,000 years ago

Reptiles (including dinosaurs)
roamed the Earth
200,000,000 years ago

Land plants and
amphibians emerged
300,000,000 years ago

The first fish
began swimming in the seas
400,000,000 years ago

Evolution Math

Look at the clock to find the answers to the questions below.

a. Life began on Earth _____

b. Jellyfish and coral first appeared in the seas _____

c. Fish began swimming the great waters of the world _____

d. Land plants grew and amphibians first crawled among them _____

e. The dinosaurs began to roam the Earth _____

f. Humankind—in the form of man-like apes—first appeared _____

g. You first appeared on Earth _____

This Time, Picture It

Draw your own version of the clock of evolution on the back of this page. Use pictures instead of words to show what was happening at particular times.

Evolution: A Time and a Place for Everything

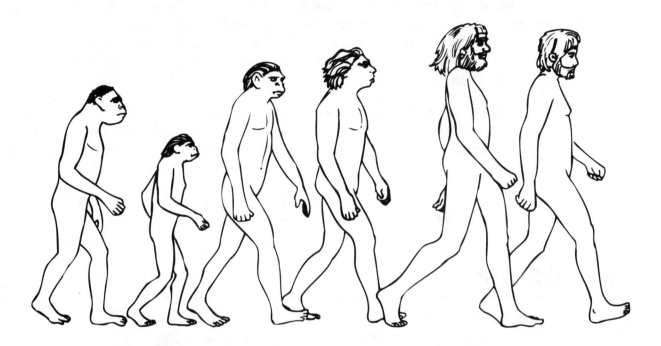

The theory of evolution was popularized by Charles Darwin, a famous naturalist who found that even animals with a common ancestry could adapt to different ways of life. He recognized that new species could arise as a result of the buildup of small changes from one generation to the next. He also realized, however, that this is a very slow process. The changes from generation to generation would be so small as to be virtually unrecognizable. It would take a creature the size of a mouse, for example, 60,000 years to evolve to a creature the size of an elephant.

Since the beginning of time, "doing the right thing at the right time" has been the most critical survival strategy of all living things. Thus, biological clocks, which were carefully synchronized with the external rhythms of nature, were given a high priority in evolution. These clocks are incredibly ancient and are found in the most primitive of species.

The human inclination to "nap" in the afternoon may be caused by the ticking of an ancient and out-dated body clock. Our ancestors probably had to nap during the intense afternoon heat. This mid-day sleep interlude is still programmed into the primitive part of our brains. This ancient urge is overridden by higher, more modern centers of thought that allow us to resist this temptation and say "no" to the old program.

Sometimes, the biological clocks that make a species tick fall out of time with nature. If the rhythms cannot be brought back together, the results can be disastrous.

Fast Fact: More species of plants and animals have died out than exist at present.

Fast Fact: Scientists think that up to present day, 99% of all the things that have ever lived on Earth have died out naturally.

Fast Fact: The ginkgo tree is the only plant that has been around since dinosaurs roamed the Earth.

Extinction Is Forever

When the number of deaths within a species is greater than the number of births, an entire species can disappear–it's called an extinction. Extinctions are and have always been a natural part of life on Earth; one species dies out to make room for another. Some species are unable to adapt to changes in the environment and they become extinct. Sometimes two species compete for the same necessities of life–one species is successful and the other dies off.

There is a relatively new reason for extinction: human beings. Human interference in the evolutionary process produces an extreme and rapid change and competition. We kill animals in such huge numbers and destroy natural habitats at such an alarming rate, that surviving creatures cannot replenish their stock.

For many species on this planet, time is running out. The number of extinctions has increased greatly in recent years. If the trend continues, we will lose 50,000 species a year by the turn of the century.

Survival Status

Five milestones have been established to help determine the survival status of plants and animals on the road to extinction:

Vulnerable: species at risk because of low or dwindling numbers (or other reasons)

Threatened: species likely to become endangered if whatever is making them vulnerable is not reversed

Endangered: species threatened with imminent *extirpation* or *extinction* throughout all or a large portion of their territory

Extirpated: species no longer found in the wild in their own country but can be found elsewhere

Extinct: species no longer found anywhere in the world

Overtime

Dedicated people around the world spend their lives trying to save species from extinction. Governments, world wildlife federations and concerned citizens have set up special wildlife parks, reserves and sanctuaries to help endangered species. Thanks to their efforts, a number of species have been "de-listed" or brought back from the edge of extinction.

Extinction Scramble

Below are the mixed-up names of a number of species that are on the road to extinction. Unscramble the letters and use the hints to discover the names if these creatures.

Threatened

1.   YNSIP   OSTF-HLELS   UTRELT   _____
      Hints: prickly   not a hard case   "withdrawn" animal
2.   LCAKB   DRHESREO   _____
      Hints: not white   colored farm animal
3.   URBORWNGI   WLO   _____
      Hints: digging   hooter
4.   NEPI   TNERMA   _____
      Hints: kind of wood   boy's name

Endangered

1.   LBEU   RCEAR   _____
      Hints: sad   fast
2.   ROAAUR   TOURT   _____
      Hints: Borealis   fish
3.   HPOGWONI   NECRA   _____
      Hints: hollering   construction apparatus
4.   LBUGEA   HLEAW   _____
      Hints: bulgy headed   biggest mammal

Extirpated

1.   GYPMY   ROSHT-NDHRE   OZDRILA   _____
      Hints: dwarf   not long horned   reptile
2.   DPAELDSHFI   _____
      Hints: gill breather with oar
3.   RETARGE   RIEPARI   NKCICEH   _____
      Hints: bigger   flat plain   "foul"
4.   SFWIT   XFO   _____
      Hints: fast   sly fellow

Extinct

1.   LUBE   LWLEAEY   _____
      Hints: sad   big eyes
2.   GEPNSASRE   GPOENI   _____
      Hints: person on board   statue-lover
3.   ASE   MKNI   _____
      Hints: ocean   coat rodent
4.   RGATE   KAU   _____
      Hints: fantastic   black and white seabird

If you haven't heard of these creatures before, look them up . . . before it's too late.

Definitely Decayed

The most important law of science and the ultimate law of nature is that all things eventually wear out-even human beings. When we talk about the life cycle of living things, decay is the last part of the cycle. Living things are born, they live, they die and they decay. Nonliving things wear out, too. Beautiful, bright, bouncy balls fade and deflate; dolls lose their hair and their stitching; tree houses slowly rot and fall down. In nature, the passage of time shows itself in both living and nonliving things as decay.

Nature's Life Cycle

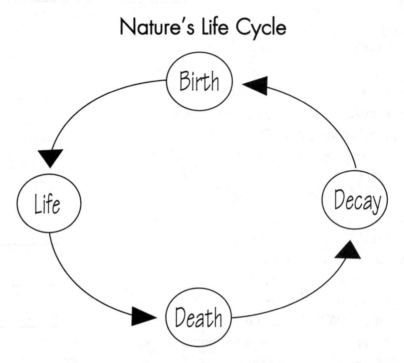

The Life Span of Our Universe; As in the Beginning, So in the End

For our universe, time began some 15,000 million years ago with what scientists call the "Big Bang"-an immensely violent explosion that brought clouds of matter together and turned them into galaxies. Our universe is still expanding, but it cannot go on forever. No one knows how or when it will end, but it will end sometime. One theory is that the universe will slow down and eventually stop expanding. When this happens, the pull of its own gravity will force it to collapse back on itself. All the processes of expansion will run backward, until . . . the "Big Crunch." After the universe is compressed into one small, but extraordinarily energetic blob, a new "Big Bang" will occur. If this theory is right, even the universe has its own rhythmic beat!

Chapter 3
24-Hour Day

The 24-Hour Day: Our Basic Unit of Time Measurement

One complete 24-hour day is based on the sun's movement across our sky which is caused by the eastward rotation of the Earth about its axis.

From the earliest days humans have marked time by a cycle of darkness and light. The sun appears to rise in the East each morning and lighten the sky, travel to reach its highest point at noon and then descend causing waning daylight before it sets beneath the western horizon and the darkness of night falls on the Earth.

Although we talk about the sun rising or setting, or moving across the sky, it is not in fact the sun that is moving-it's the Earth. The Earth is like a giant ball turning on what is called its axis-an imaginary pin that runs through the center of the Earth between the North and South Poles. The Earth rotates, or spins, once every 24 hours making the sun appear to rise in the east and set in the west.

Some people call the Earth itself a timekeeper, whose rotation makes up our basic unit of time measurement-our 24-hour day.

Both sides of the Earth experience the cycle of day and night-but not at the same time. The part of the Earth that is facing the sun experiences daylight and day. The part of the Earth facing away from the sun experiences darkness and night. When the sun peeks over the horizon, day begins for you, but the sun sets on children on the other side of the world who wear their pajamas and settle into bed. As the sun rises on one side of the Earth, it sets on the other.

Fast Fact: Although we cannot feel the motion, the Earth is constantly spinning on its axis. At the North and South Poles, it moves very slowly, but on the equator it is speeding at 1,000 miles an hour!

How Long Does Daylight Last?

The length of time we have daylight and darkness each day varies according to the seasons, when the Earth tilts at different angles in relation to the sun. In winter we have long nights and short days, and in the summer we have long days and short nights. At the North and South Poles there are times when the sun doesn't rise at all! The only time we experience 12 equal hours of light and darkness are at the spring and winter equinoxes.

Daylight Math

Your local newspaper will tell you what time the sun rises and sets on a particular day. Have one child record the stats for the day's sunrise and sunset. Have a group figure out how much daylight there will be each day.

Fast Fact: Ancient Egyptians divided day and night into 12 equal parts and altered the length of the evening hours to compensate for the seasonal differences.

Name _____

The Cycle of Day and Night

Illustrate and color the pages of this booklet. Cut out each page and then staple your booklet together.

Staple here.

The Cycle of Day and Night

by _____ 1

Staple here.

Each day the sun rises in the east and a new day begins. 2

Staple here.

The sun rises until it reaches its highest point in the sky-at noon. 3

Staple here.

After noon the sun begins to sink in the sky. 4

Staple here.

The sun dips lower and lower until its sets in the west. 5

Staple here.

The sun's light disappears and dark night falls. 6

Staple here.

Tomorrow the sun will rise again! 7

Staple here.

In the day I like to _____.
At night I like to _____. 8

Our Spinning Earth

Materials

globe
lamp or flashlight
tiny human figure
Blutak™ or plasticine

Ask children to think about day and night as discussed above. Explain that the globe is a representation of the Earth. Show how it spins on its axis. Have children pretend they are the Earth and spin around once. *Does anyone know how long it takes the Earth to make one complete revolution?* Explain that the light represents the sun and the figure represents "us." Have children help to pinpoint your location on the globe. Use the tac or plasticine to help fasten the figure to the globe on your part of your country.

What to Do

1 Ask children to picture themselves on the Earth, spinning, even though they cannot feel they are moving.

2 Shine the light on the globe.

3 Turn the globe so the figure is in the dark, halfway around the globe from the direct light beam. What time of day are you having at this time? It's midnight. The end of one day and the beginning of the next.

4 Turn the globe so the light shines directly on the figure. What time is it now? Day is correct–noon to be exact. It is noon when the Earth has turned halfway around in its rotation. One half of the day has passed.

5 Turn the globe back to simulate a sunset position when three quarters of the day has passed.

6 Turn the globe back to the midnight position to show the passing of one complete day–one rotation.

7 In turn, have students to come up to the globe to show a position for: day, night, sunrise, sunset, midnight, noon, rotate the globe to show one complete day, one half of a day and so on.

Try This

- Young children can best relate to the terms *light time*, *dark time*, and *day* and *night*.

- **Talk about day and talk about night.**

- What do you do in the day? What do people you know do? What do animals do? What do you do at night? What do animals do? Do you know of any people who are awake all night?

The Cycle of Day and Light

You can't feel it, but our Earth is always spinning on its axis. This turning of the Earth causes day and night. It makes one complete rotation every 24 hours–giving us our 24-hour day. In this picture of the Earth and sun, one side of the Earth is facing the sun. The sun lights up this side of the Earth and it's daytime here. The other side is turned away from the sun. It's dark and shaded. It's nighttime here.

Color the day times yellow and the night times black.

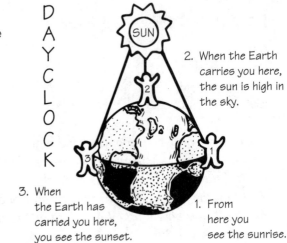

DAYCLOCK

2. When the Earth carries you here, the sun is high in the sky.

3. When the Earth has carried you here, you see the sunset.

1. From here you see the sunrise.

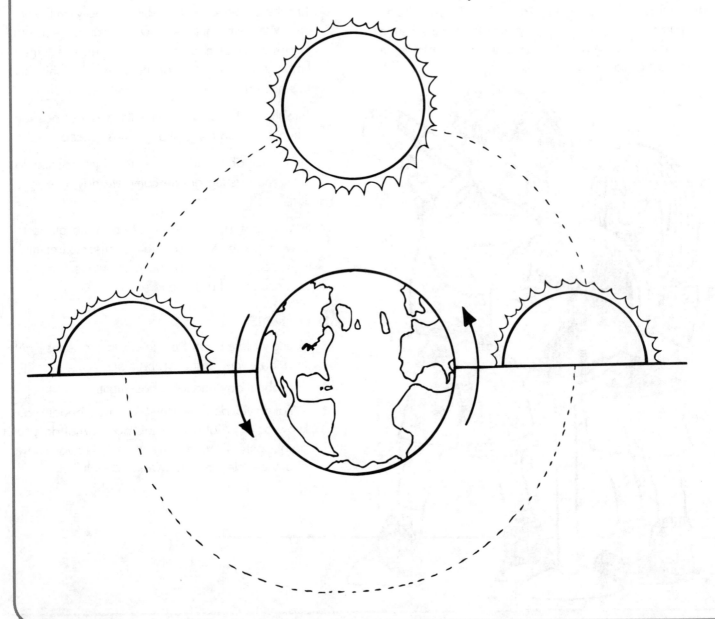

24-Hour Days

In early times, people marked the day by sunrise and sunset. As life became more complex, more times were needed. The Romans identified a third time when the sun was halfway between sunrise and sunset – as noon. By A.D. 605 the Roman day was divided into seven periods or canonical hours which were used throughout Europe for many centuries.

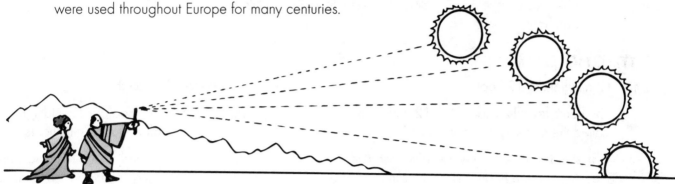

The 24-hour day is based upon the natural rotation of the Earth. About 5,000 years ago the Babylonian civilizations developed the 24-hour system using their base 6 counting system. This system explained the sun's 360 degree circle across the sky, approximated 360 days in a year, with 12 months and 30 days to a month. The days were broken down into 6 periods, 12 periods and then 24 as the units were needed.

There are _____ hours in one day, _____ minutes in one hour and _____ seconds in one minute.

These are man-made units based on _____

our largest whole unit of time measurement is the _____

Why do you think that some people call the Earth a timekeeper? _____

_____

_____

_____

_____

How many minutes are there in two hours? _____

How many seconds are there in one hour? _____

Challenge

How many minutes does it take for the Earth to rotate once on its axis? _____

Twice Around the Clock; a.m. and p.m. Hours

Each day begins and ends precisely at midnight and is divided into two equal 12-hour segments. Most clock dials show the 12 hours. The hour hand makes two complete turns of the clock face in 24 hours.

When we use the 12-hour clock, times are followed by *a.m.* or *p.m.* to tell us if it is a morning or afternoon hour.

a.m. and p.m.

a.m. = 12 midnight to 12 noon

The hours from 12 midnight to 12 noon are called the morning or a.m. hours.

a.m. stands for the Latin words *ante meridiem*, meaning "before noon–before the sun is on your meridian line."

p.m. = 12 noon to 12 midnight

The hours from 12 noon to 12 midnight are called the afternoon or p.m. hours.

The letters *p.m.* stand for the Latin words *post meridiem*, meaning "after noon–after the sun is on your meridian line."

Circle *a.m.* or *p.m.* under each picture.

a.m. *or* **p.m.**

a.m. *or* **p.m.**

a.m. *or* **p.m.**

a.m. *or* **p.m.**

a.m. *or* **p.m.**

a.m. *or* **p.m.**

Is 9:00 a.m. in the morning or nighttime? _____

At 12:00 a.m. might you be sleeping or eating your lunch? _____

The Sun as Timekeeper

The sun began to light up the Earth about $4\frac{1}{2}$ billion years ago and as long as people have lived on Earth they have organized their lives around the sun. Many ancient people measured the time by watching the position of the sun and marking its movements across the sky. Native people of North America, Celtic people of Europe and the Mayans of South America all relied on the sun as their time-keeper. Structures of stone and earth were built in some places to help early people keep track of the days, months, seasons and years. Some were constructed to frame the setting sun on the day of winter or summer solstices. In Saskatchewan, Canada, at least two prehistoric rock configurations have been discovered dating from about the first century A.D. and the most famous of such timekeepers, Stonehenge, is found on the Salisbury Plains of England.

The sun gave people their first means of timekeeping. It allowed people to track the seasons and the years and to mark the days and the hours. Its position in the sky and the shadows the sun cast gave people their first means of timekeeping. The Egyptians were the first to use the sun as a timekeeper in this way.

At first people relied on shadows of themselves and natural objects to read the time of day but later turned to shadow clocks and sundials for this information.

Fast Fact: The turning of the Earth is a dependable clock that serves as the basis for our time-keeping practices.

Estimating the Time

Morning or Afternoon?
The sun rises in the east and sets in the west. If you know your compass directions, you can tell whether it's morning or afternoon by following the sun's movements.

Noon
You can use your judgment to tell when the sun has reached its highest point in the sky. This is your rough estimation of noon time.

How Much Time 'Til Sundown?
When you have determined that the sun is setting and not rising, you can estimate the time of day. Hold your arm at arm's length and face the sun. Raise your thumb, bend your wrist and position your hand between the sun and horizon. Allot 15 minutes for each finger that fits in your visual range between the sun and the horizon. This will tell you how much time is left before sunset. If four fingers fit in this range, there are 20 minutes to sundown.

Your Shadow Knows

The sun is most intense at particular times of the day. Your shadow is at its shortest when the sun is at its strongest. Teach children to look at their shadow for indications of time and "skin safe" information. The sun's harmful rays are strongest at noon.

Shadow Time

Early people relied upon shadows to tell time throughout the day. After taking part in this activity, children will see that they, too, can rely on the sun and their own shadow to tell time.

Materials

sunny day
open space in an undisturbed place
chalk or washable paint and brush and pavement
sandy ground
stick

Get Started

Have children find partners. Distribute the marking materials needed for your area.

How do you think a shadow can help to tell time? Are shadows always of the same length? You can tell the position of the sun in the sky by looking at the length and position of a shadow.

What to Do

1 Go to your sunny place first thing in the morning.

2 Each pair finds their own space and draws an X on the ground.

3 One partner stands on the X while the other traces the shadow that is cast and records the time of day beside the shadow.

4 Return to this spot every hour to make another shadow tracing.

5 Talk about changes in the shadow. Draw children's attention to the movement of the sun and its affect on the shadows. Can children predict the next location of the shadow? What happens to the shadow at noon?

Try This

- You will find that the morning shadows are long and cast towards the west. The noon shadow is the shortest shadow, and the afternoon shadows are cast towards the east and get longer as the time passes.

- **High Noon**
  According to the sun, our shadow is the shortest at high noon. Check your clock and your shadow at noon. Twelve noon on your clock may not occur at high noon sun time in your area, because of standardized time. Because of standardized time, the solar and clock times sometimes do not agree. Your shadow time may not match your clock time.

- This activity works just as well using a stick in place of the live body!

Step on your shadow and you'll have good luck!

Sundial

Long before wristwatches were vogue, people turned to the shadows to help them tell time. As the Earth rotates, the size and direction of shadows changes. Early people relied on these shadows and human estimations based on the sun to help them tell time. Eventually the shadow watchers invented the sundial– a kind of shadow clock. It consists of a circular face marked with numbers to represent different times. A pointer, called a *gnomon*, sits in the center to cast its shadow across the circular face to indicate the times of day.

Materials

8" x 12" (21 x 30 cm) piece of paper
plasticine or clay
4¹/₂" (11.5 cm) pencil, straw or wood dowel
masking tape

Get Started

Borrow a garden version of a sundial to help introduce this topic. Explain that the shadow cast by the gnomon or needle shifts as the sun moves through the sky from sunrise to sunset. This shadow indicates the time of day. Have you ever seen a sundial? Where might you find a sundial?

What to Do

1 Tape the paper to a sturdy board, book or other solid surface.

2 Mark an X 6" (15 cm) from the ends and 2" (5 cm) from one side of the paper.

3 Stick the clay in this spot.

4 Find the center 6" (15 cm) mark on the opposite edge of the paper and mark a tiny sun here.

5 Use your pencil to connect the sun and the X.

6 Poke the gnomon into the clay with a very small slant towards the "sun."

7 Just before 12 noon, take your sundial into the sun. Do not move it from this place once you begin.

8 At high noon (or 1 p.m. daylight saving time) the shadow should be cast along the line you have drawn towards your "sun picture." Adjust your gnomon until the shadow falls in this way.

9 Pencil in a line to indicate the position of the shadow every hour on the hour. Write the standard time indicated by your watch on each line.

Quick Query: Would a sundial be useful every day? What conditions must be present?

Fast Fact: Early Saxons used wall-mounted, stone sundials to indicate the time. The day was divided into "tides" for church services.

Fast Fact: Each degree of longitude passes in front of the sun for four minutes each day.

World Standard Time

Because the Earth is always revolving, the sun's position in the sky is different wherever you are. Not long ago, every town in North America relied on its own local time set by some form of this sun time.

As railroads and people traversed the countryside, life became more time-dependent and the need for an international system of time measurement grew. The concept gained acceptance in 1880, when Britain began using London time as a standard time. In 1883 the railroad established a standard Railroad Time and in 1884, representatives from around the world met in Washington, D.C., to discuss methods to simplify timekeeping.

International representatives agreed to divide the world from North to South into 24 almost equal sections, or time zones, by 24 meridians. Most maps and globes contain these imaginary meridians based on lines of longitude. There is one time zone for each hour of the day and all places within a zone share the same time. Each zone is about 1,000 miles wide and covers 15 degrees of longitude–the distance the Earth turns in one hour.

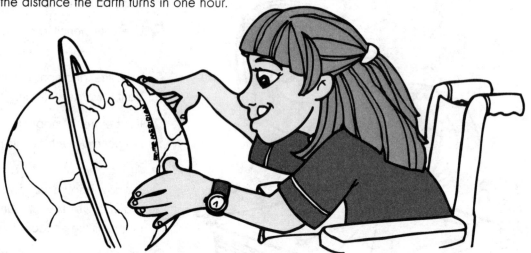

Look for the meridians on your map or globe. You will see that adjustments were made to accommodate oceans, mountains and irregular country borders.

Time measurement begins at the prime meridian, the line of 0 degrees longitude, which runs through Greenwich, England. We call time in this zone Greenwich Mean Time (GMT). All other times are based on GMT. When you travel east of the prime meridian, you add one hour for every time zone passed and when you travel west, you subtract one hour for every time zone passed.

Time based on this system is called Greenwich Mean Time (GMT), World Standard Time (WS) or Universal Time (UT). This system was adopted by most countries of the world. It eliminates time differences in any one area, ensures that 12 noon occurs at, or close to high noon everywhere and makes it easier to coordinate international events.

The continental US. stretches so far east and west that it contains four time zones; Pacific, Mountain, Eastern and Central Time.

Fast Fact: On March 19, 1918, 35 years after the clocks were standardized, the U.S. Congress passed the Standard Time Act which adopted WS as the official time of the nation.

Fast Fact: In 1964, Coordinated Universal Time, based on atomic time and the solar year, replaced GMT as the universal standard.

World Standard Time

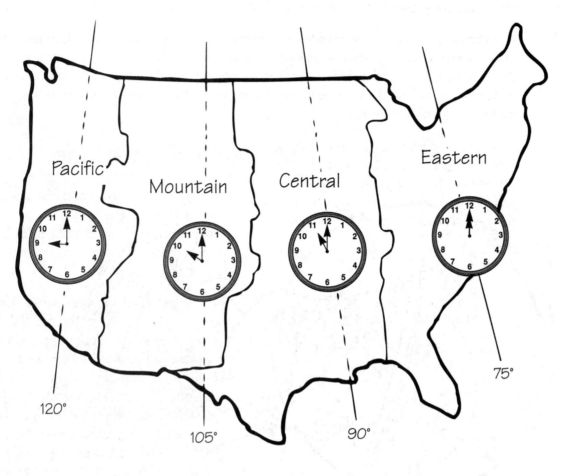

Pacific

Mountain

Central

Eastern

120°

105°

90°

75°

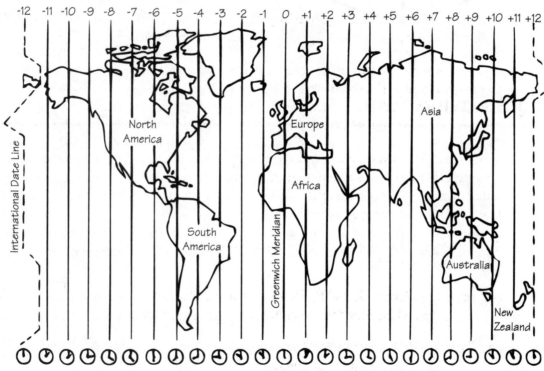

-12 -11 -10 -9 -8 -7 -6 -5 -4 -3 -2 -1 0 +1 +2 +3 +4 +5 +6 +7 +8 +9 +10 +11 +12

International Date Line

North America

Europe

Asia

Africa

Greenwich Meridian

South America

Australia

New Zealand

Traveling Through Time

If you travel north or south along a line of longitude, you remain in the same time zone your entire journey. *Follow a line of longitude from the North Pole to the South Pole on the globe or map to demonstrate.*

If you travel from east to west along a line of latitude, you cross into new times zones, and in a sense, travel through time. *Follow a line of latitude to demonstrate.* Travelers who cross from one time zone into another must adjust their watches one hour for every 15 degrees of longitude crossed. Travelers sometimes rely on this saying to help keep the time straight: "Go west to the new day; Go back east to the old."

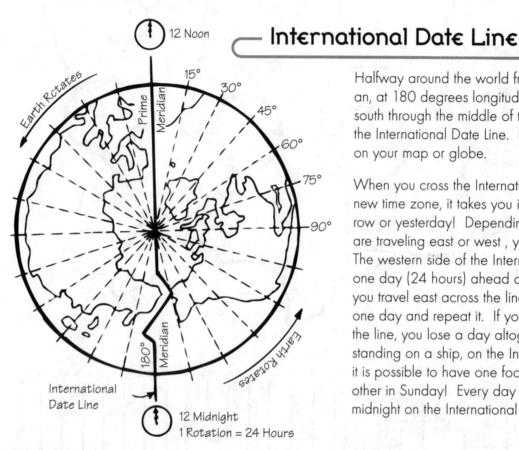

International Date Line

Halfway around the world from the prime meridian, at 180 degrees longitude, running north to south through the middle of the Pacific Ocean lies the International Date Line. Find this imaginary line on your map or globe.

When you cross the International Date Line into a new time zone, it takes you into a new day—tomorrow or yesterday! Depending upon whether you are traveling east or west, you lose or gain a day. The western side of the International Date Line is one day (24 hours) ahead of the eastern side. If you travel east across the line, you would go back one day and repeat it. If you travel west across the line, you lose a day altogether. If you are standing on a ship, on the International Date Line, it is possible to have one foot in Saturday and the other in Sunday! Every day begins and ends at midnight on the International Date Line.

Daylight Saving Time

Daylight Saving Time is a method for adjusting regional standard time to take advantage of the most hours of daylight. The concept was presented by Benjamin Franklin in 1784 and used during the First and Second World Wars in North America. In the 1940s, after the Second World War, most of North America adopted the system.

In places where Daylight Saving Time is used, the clocks "spring" ahead by one hour on the last Sunday in April and "fall" back in the fall on the last Sunday in October at the official hour of 2:00 a.m.

Quick Queries
• *What are the advantages to having more waking hours of daylight?*
• *Do most people get up at 2:00 a.m. to change their clocks? How do we get around this early morning clock change and still be on the correct time?*
• *If you forget to change your clock in April, would you be early or late for school on Monday morning? What if you forgot in October?*

World Standard Time

World Map

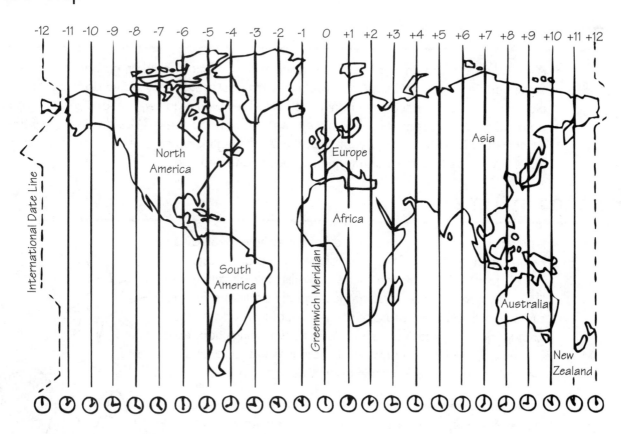

Fill in the compass (N, S, E, W).

Trace the prime meridian with your pencil.

1. There are _____ ° of longitude and _____ meridian lines. These imaginary lines run (circle the correct answer) north and south/east and west on the map or globe.

2. The line that passes through Greenwich, England, is called the _____.

If it is 12 o'clock noon, Monday, in Greenwich,

3. What time is it 15° east of Greenwich? _____

4. What time is it 15° west of Greenwich? _____

5. What time and day is it 180° east of Greenwich? _____

6. What time and day is it 180° west of Greenwich? _____

7. The 180° meridian is also called the _____ because _____

_____

Chapter 4
Measuring Time

Imagine

Ask children to come on a journey back in time with you. Have children go back to a time long ago when people lived in nature. *What would they do each day?* They would do many things to ensure that they could exist from one season to the next. They would gather, hunt and plant foods and prepare for the coming seasons. *Would they need to know the time? Would knowing what day it was, what season it was or what year be important?* Although knowing precise times wasn't always important, knowing general times of day, month, season and year was.

Humans in Time

As long as people have lived on the Earth they have been aware of the passage of time. They knew of the rhythms of the sun, moon, stars, weather and the cycles of plants, animals and human life. Early people understood the passage of time from sunrise to sunset, month to month, season to season, year to year and youth to old, age. They didn't always understand why things happened as they did, but they were aware of cyclical events which told them what they needed to know about time.

At first people reasoned that the heavenly objects revolved around the Earth in a repetitive course but later realized that the cycles were based on the movements of the Earth and other heavenly bodies around the sun in measurable orbits.

They measured time by the solar day–the time it took the Earth to rotate once completely and knew that its position in the sky indicated the time of day. They measured the solar year of 365$\frac{1}{4}$ days–the time it took for Earth to complete its orbit around the sun. They knew that the position of the sun determined the seasons and its reappearance in an identical position in the sky indicated the passage of one year. They measured the lunar month, the time it takes for the moon to orbit the Earth once, to be 29$\frac{1}{2}$ days.

Quick Queries: Why were early people so aware of the natural cycles?

Why Did People Start Measuring Time?

Early people relied on nature to provide them with the necessities of life. They were dependent on agriculture, wild vegetation and game for their survival. Their survival depended upon knowing when to plant, hunt, harvest and gather. Farmers, food gatherers and hunters of early civilizations developed methods to measure the daily and seasonal cycles of the natural world that indicated change, growth and renewal.

Why Do We Need to Know the Time?

In the past people's life-styles allowed for estimates of the time. As life became more complicated, more accurate means of time measurement became necessary. Soon the day was divided into smaller and smaller units of time. Today we have standardized times and need to know the right date and minute or just about everything or we'll miss out. Our schools, hours of employment, means of transportation, sports events, health care and entertainment have become entwined with precise dates and times. Our society runs on a tight time schedule!

Quick Queries:
- *Have you ever been late or have the incorrect date? What happened?*
- *Is your life controlled by the clock and calendar?*
- *Do we really need to live with such specific times? What effects do these rigid time constraints have on you? On society? Can you think of ways that more flexible times might fit into society?*

How Do We Measure Time?

From ancient times until now the sun has always been the basis of our time-keeping practices. Time can be measured by just about anything that moves at a consistent rate.

Throughout the ages people have developed various methods to keep track of and define quantities of time – at first by the consistent cycles of the sun, moon, seasons and the solar year and then by sundials and hour-glasses. Today we turn to mechanical time-keeping devices of great accuracy to measure the moments for us.

Official Time

The Official Time is coordinated and provided by the International Committee of Weights and Measures. The official U.S. Bureau of Standard's talking clock timepiece is relied on by many.

Exact time signals are given by the Naval Observatory and sent out to various parts of the world.

Measuring Time
Ticks and tocks
and rhythms and clocks
mark and measure the moments.

Explore Time

Materials

sand and water table or large basin filled with each
4 funnels of various sizes with plugs for each group,
    (can use homemade funnels made from plastic jugs
    with the bottoms removed and caps retained)
buckets or large cups to fill the containers
timer or stopwatch (optional)
reproducible on page 53

Get Started

Pair the children into groups of two to four.
Have the groups label four funnels with the numbers 1-4

What to Do

1 Provide children with a means to time the flow of material through the funnel. Counting, stopwatches or timers will work. Instruct children how to properly use this method. Can the children think of ways to reduce this margin of error?

2 Children proceed to the water table and look at their funnels. Which one do they think will hold the most water?

3 Children fill funnel 1 with water and think about how long it will take for the water to run out of the particular funnel. Each child should make a verbal guess.

4 Children remove the plug and start the timing device. Timing is stopped when the flow of water stops.

5 How long did it take for the water to run out? Have students record the results for each container on the reproducible provided.

6 Children repeat the process with water for all four funnels. Why did some funnels take longer than others to drain? Look at the shape and size of the funnels.

7 On the same or another day, have children proceed to the sand table. Do they think it will take more or less time for the sand to run through the funnels? Why do they think this? Record predictions.

8 Children use the same funnels and sand in the place of water to repeat the process outlined above.

Funnel Time

Time the water and sand flowing through your funnels. Record your results on this page and answer the questions below.

Water

How much time did it take for the water to flow through each of the funnels. Record your results below.

#1 _____   #2 _____   #3 _____   #4 _____

Sand

Do you think it will take the sand more or less time to flow through the funnels than the water?

Why do you think this?

_____

_____

_____

Record your results below.

#1 _____   #2 _____   #3 _____   #4 _____

It took the most time for the _____ to run out of funnel _____.

It took the least amount of time for the _____ to run out of funnel _____.

Time Comparison Centers

1-4 • Math • Exploring Intervals/Manipulating Materials/Understanding Relationships/Estimating/Recording • 60 min

Materials

Center A: Cups of water with ice cubes
Center B: 2 wind-up toys
Center C: 2 sand timers
Center D: 2 marble mazes and marbles
Center E: 2 pendulums

Get Started

1 Set up five centers as indicated above for the children to visit.

2 Demonstrate each center activity for the children. Show the children how to record their guesses about which item will take longer to complete the task.

3 Ask children to think about which actions will take the longest.

4 Explain that each group is permitted five minutes at each center. The beginning and end of each center session will be indicated by the timer. Let them hear it.

5 Divide the class into eight equal groups for exploration of the centers.

6 Hand out one record sheet for each group and explain the rotation system for the activity.

What to Do

1 Have children proceed to their first center and signal them to begin.

2 Observe as children explore time interval comparisons in their own way. Introduce concepts and vocabulary, duration, shorter duration, longer duration, interval, faster and slower.

3 Have children record their findings.

54

Time Terms and Units of Time Measurement

There seems to be a whole language associated with time and the telling of time!

Write the letter of the matching description in the blank.

1. _____ units

2. _____ day

3. _____ hour

4. _____ minute

5. _____ second

6. _____ morning

7. _____ noon

8. _____ afternoon

9. _____ evening

10. _____ night

11. _____ midnight

12. _____ month

13. _____ solar year

14. _____ decade

15. _____ century

16. _____ millennium

17. _____ seasons

18. _____ clock

19. _____ watch

20. _____ timer

21. _____ calendar

A. measures intervals of time but doesn't tell the time

B. a man-made unit of time measurement; 60 minutes

C. unit of 1,000 years

D. a small man-made unit of time; $1/60$ of a minute

E. a unit of 60 seconds

F. natural and man-made measurements of time

G. the time of day from sunset to sunrise

H. time it takes for the moon to orbit the Earth

I. daylight is waning but it's not yet dark

J. the exact middle of the night, 12 a.m.

K. a unit of 10 years

L. instrument that measures hours, minutes and seconds

M. the part of the day between sunrise and noon; a.m.

N. a small clock that you can carry with you

O. 24-hour time period it takes the Earth to rotate once

P. unit of 100 years

Q. a system we use to mark the days and years

R. the time between 12 noon and sunset

S. the exact middle of the day, 12 p.m.

T. a natural unit of time that equals 365.26 days

U. quarters of the solar year with varying weather, hours of daylight and effect on living things

Measuring an Instant

The length of time that something continues, or lasts is called the duration. Can we measure duration?

Well, let's start with a very small duration of time–like an instant–can it be measured? Stamp your foot, clap your hands, click your pencil and think of ways that you can measure how long this action takes. We can measure these with ticks or small instants on a clock. The slight sharp sound a clock makes is called a tick, and it marks a small quantity of time called a second.

Just a Second

Just a second, before we go any further, how long is a second? That is an easy question to answer, a second is very short, in fact it is the smallest whole unit of time in our system of time measurement. It is $1/60$ of a minute–there are 60 seconds in a minute. Count "One Mississippi," clap once, hop once, blink your eyes slowly, jump rope once–these actions all take about one second. *What else takes one second?*

Fast Fact: An exact second is called an SI Second. It is based on a particular number of vibrations in a specific kind of cesium atom. This second can be broken into microseconds (millionths of a second) or nanoseconds, (billionths of a second) and is the foundation of Coordinated Universal Time.

Wait a Minute

Now wait a minute; how long is a minute? It's 60 seconds. Count to 60 and you have one minute. It takes 60 minutes to make one hour.

Just a Cotton Pickin' Minute

K-4 • Science/Math • Time Measurement/Estimation/Counting • 5 min

Materials
cup and six cotton balls for each participant
timing device

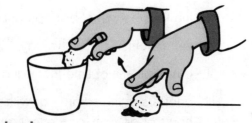

Get Started
Provide each child with one cup and six cotton balls.

What to Do

1 Instruct participants to measure one minute by "picking cotton." There are six cotton balls, one should be dropped into the cup every 10 seconds.

2 On a signal, begin timing while participants begin "picking."

3 Children raise their cup when it is full and stop "picking" when the one-minute signal has been given. Were the timers accurate? Too fast? Too slow? Does the timing improve with practice?

How Much Time Does It Take?

2-4 • Math/Physical Education/Music • Understanding Time Passage/Time Measurement/Counting/Comparing/Recording/Handling a Time-Keeping Device • 30 min

Help children to understand the passage of time and the measurement of time intervals

Materials

as needed for chosen activity
as needed for method of timing

Get Started

1 **Choose one action to time.** Clap your hands together as a group 10 times, sing the national anthem at the usual pace, each child in the group say their name in turn, pass a beanbag around a circle, bounce and catch a ball 25 times, jump rope 50 times, walk around the room and return to original places, do the bunny hop from here to there, say the alphabet.

2 **Choose one consistent method of timing.** sand timer, second hand on the clock, stopwatch, swinging pendulum, metronome, marble maze . . .

3 Choose one child for "timekeeper" and instruct her how to begin timing and end timing on cue.

What to Do

1 Ask the simple question "How much time will it take to . . . ?" using various actions and time-keeping methods.

2 Demonstrate the action and the timing method and prepare participants to begin timing and taking part in the action–on cue.

3 Provide the signal to begin and end the timing and the action.

4 Keep an ongoing record of the intervals taken to perform designated actions.

It's Speedy by a Nose!

Draw children's attention to professional methods of timing races. Stopwatches and digital watches count time in seconds and fractions of seconds.

Homemade Timekeepers

There are many ways to measure the time intervals. Can you think of innovative ways to time particular events? Try making a timekeeping device of your own.

Measuring Time in Music

Music is made through a combination of harmony, melody and rhythm—the organized sequence of musical beats. Music has a time and language all its own. Written music, called musical notation, tells musicians what notes to play (and how long to play them). Words are also used to provide more information about the playing of the music.

Timing is very important in music. Listen carefully to a piece of music. What do you hear? Are there sounds all through the music? Do you think the silences play a part in the music? Try singing some familiar tunes without the pauses and silences between words and beats. You will discover that sounds and silences are equally important in the making of music. The duration of musical sounds are indicated by notes and the duration of silences by rests. Time values for notes and rests range from whole units to sixty-fourths.

Basic Notes and Rests

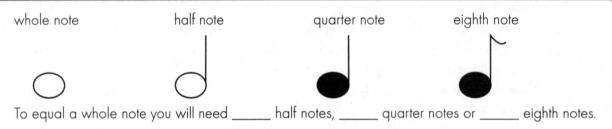

whole note   half note   quarter note   eighth note

To equal a whole note you will need _____ half notes, _____ quarter notes or _____ eighth notes.

Try This: Create math lessons based on the whole, quarter, half and eighth notes. It makes fractions relevant and leads to a better understanding of music we hear every day!

Tempo

Tempo is the rate of speed of a musical piece. It is described in Italian terms.

Common Tempo Terms
grave: solemn, slow tempo
adagio: slow, stately
largo: very slow tempo
andante: medium, graceful
moderato: moderate, smooth
allegro: quick, full of life
vivace: fast
presto: very lively and quick

Try This: Move, draw and rest while listening to music with various tempos. Try to identify the various tempos.

Fast Fact: A metronome is sometimes used to help musicians play at the correct tempo. Demonstrate with a real metronome.

Name _____

Measuring Time

Circle things that measure time in this picture.

Time Is Relevant

In one second I can _____

It takes one minute for _____

It takes one quarter of an hour for _____

It takes one half of an hour for _____

It takes one hour for _____

It takes one day (24 hours) for _____

It takes seven days for _____

In one month _____

In one year _____

I sleep _____ each night?

I spend _____ in school each day?

I watch _____ hours of television in one day?

There is _____ between one birthday and the next?

I have been alive for _____

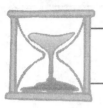

Chapter 5
Time and Technology

Time, Technology and Mechanical Timekeepers

Some people seem to have a "good sense of time" – they just seem to know how long a minute or an hour or three hours is. Most of us don't have this ability and need to rely on time-keeping devices if we want to know the time.

Throughout the ages people developed various ways to keep track of and define quantities of time. Long ago the sun provided the first means of time measurement. Today, most of the world relies on sophisticated 24-hour clocks set by world standard time to keep track of the hours, minutes and seconds and the Gregorian calendar to keep track of the years, days and months. Most countries around the world share these common systems of time measurement.

Fast Fact: Do our most accurate clocks measure the true solar day? No, they don't! Some solar days are longer than others so our most precise time-keeping devices are more consistent than the Earth's rotation!

Early Timekeepers

Ball Clock
In the 16th century Galileo invented a Rolling Ball Clock which kept time by rolling balls down an inclined surface. This idea was further developed but never really caught on.

Use a inclined marble maze of any sort as a timing device. You can measure events using "ball time" or convert to standard time by first timing the roll of the marble from point A to point B and then converting the "ball time" to standard time.

The Clepsydra
Have you ever heard of a clepsydra? This form of timekeeper was in use for a long time – telling time in its own watery way. It was a kind of water clock that measured intervals of time by the flow of water. The device was turned upside down to start water flowing from one chamber to another through a funnelling system. The accuracy was limited by changes in the rate of flow due to inconsistent air pressure and the changing water level.

Make your own water clock! Find a way to allow water to drip through a small hole in one container into another. How long does it take for all of the water to pass through? Can you make a device that will measure 10 seconds, one minute or even one hour?

Fast Fact: One of the most famous clocks is known as Big Ben. This tower clock was named after the giant bell that rings out the hours. It is found on the Victoria Tower of the Houses of Parliament in London, England.

Design a time-keeping device of your own. Discuss possibilities and then draw a design of a device that keeps time at a steady rate.

Candle Clock

Caution: The candle should be placed in a sturdy holder, in a safe location, constantly monitored by an adult.

Candles or incense sticks that burned at the same steady rate were once used to measure the passage of time. The candles or sticks were marked to indicate one-hour intervals. What problems might be encountered with this time-keeping device? Is it possible to have two candles that are exactly the same? How is flame affected by drafts? Can you leave a flame burning without supervision?

Materials

2 identical candles with the same width at the top and
    bottom (not the tapered kind)
timing device
ruler, measuring ribbon, weigh scale or balance
marker
matches

Get Started

1 Review rules of fire safety and emphasize the danger associated with matches and fire.

2 Measure the length and circumference and weight of various candles. Compare your measurements.

3 Find two "identical" candles.

What to Do

1 Burn one of the candles for 15 minutes.

2 Compare the candle to one that was not burned. How much of the candle wax burned in that time frame? Measure with a ruler.

3 Mark the candle in intervals of 15 minutes based on the measurement of burned wax provided by the first candle.

4 Note how many markings are on a candle when you begin. If you want to measure one hour, how many markings will melt away?

5 Burn one candle at the front of the room in a safe place to demonstrate. How accurate was your method?

Hourglasses and Sand Timers

3-4 • Math/Science/History/Art • Understanding of Time Concepts/Problem Solving/Manipulation of Tools • 60 min

Hourglasses were once very popular measuring devices. They consisted of a closed glass shape with a narrow waist through which sand, water or mercury flowed from the upper to the lower portion of the vessel at a steady rate. When the device was turned over, it took a period of one hour for the material to flow from the top to the bottom chamber—that, of course, is why they were called hourglasses. Demonstrate with a real hourglass.

Sand timers, popular variations of hourglasses, were first used in Germany in the 17th century. These sealed vessels contained sand which flowed as described above. Sand timers were often made to measure one hour but could be made to measure other periods of time. The rate of flow depends upon the amount of sand contained and the size of the waist or hole through which the sand flowed. From ship watches to cooking, sand timers have been reliably used to measure the flow of time for many years.

Materials

tape
2 dry identical plastic soda bottles
dry sifted or sifted or colored sand
washer

What to Do

1 Fill one of the bottles about half full of sifted sand.

2 Place the washer on the neck of this bottle.

3 Place the empty bottle, upside down, so the necks of both bottles meet and the regulator is sandwiched in between.

4 Tape the necks of the two bottles securely together.

5 Turn the sand timer upside down to get the sand flowing.

6 Use another timer to find out how long it takes for the sand to run through your device. What can be done in that time frame? Time some activities using your sand timer.

Try This

*   Alter the amount of sand in your bottles to measure a specific amount of time—1 minute, 5 minutes, one hour or whatever unit you choose.

*   Set up a display of various hourglasses and sand timers. Have children determine the period of time measured by each. What is each timer used for?

Quick Query: *What is the difference between a sand timer and an hourglass?*

Early Mechanical Timekeepers

Clock mechanisms, may in fact be the first complicated, commonly used, mechanisms ever invented. There have been a great many timekeeping mechanisms developed throughout modern history to make hands move around a clock face at a consistent speed to keep pace with the 24-hour day. Throughout time these devices have become more and more accurate and capable of measuring smaller and smaller units.

About a century ago, these mechanical timepieces became very popular even though the sun remained the official timekeeper by which these clocks and watches were set.
Think about how clocks and watches are set today.

The first mechanical system of marking time was possibly a system of bells used in 996 by Pope Sylvester II–which may explain the origin of the word *clock*–from the German word *glocke*, which means "bell."

By the 13th century, the first true mechanical clocks were in use. These clocks were more accurate than earlier time-keeping devices and had no need of sand, water, sunlight or constant supervision. They were powered by slowly falling weights that were attached to gear trains. These mechanical clocks told the time but not very accurately.

Fast Fact: The oldest existing mechanical clock was made in 1386 and can be found in Salisbury, England.

The Pendulum Clock

In 1583, the Italian astronomer-scientist Galileo timed a swinging chandelier and made observations that led to the scientific principle of the pendulum. A pendulum is a swinging weight suspended from the end of a string or rod. Galileo discovered that the period taken for one swing is dependent, not on the weight at the end, but on the length of the pendulum. The "beat" is decreased by one-half if the length of the pendulum is decreased one-quarter and sequential beats take the same amount of time whether the swing is short or long.

Galileo's discovery led to the development of the pendulum clock—the first precise mechanical means of telling time. A working clock wasn't made until 1657 when Christiaan Huygens linked the regular swing of the pendulum to the movements of the clock mechanisms and made the first pendulum clock. In 1666, Robert Hooke refined the design and built the first accurate pendulum powered clock.

The pendulum clock consists of a face, a swinging string or rod with a weight (or bob), a gear train which powers the hand or hands of the clock to keep pace with the sun. Friction which slows the swinging pendulum is compensated for through the use of levers, an escape wheel and a slowly falling weight which provides the energy that keeps the pendulum in motion. A pendulum can be made to swing at a regular rate, and the length of the pendulum can be adjusted to exact seconds.

Pendulum clocks were first built in high towers that provided adequate distance for weights to fall. By 1680 a minute hand had been added to the clock face and soon after the second hand. Although these early clocks were considered accurate at the time—they could be off by two hours each day! Over time the pendulum clock was refined and by the 1900s it was accurate within $1/100$ of a second per day. People came to rely on accurate time measurements provided by the pendulum clock.

Quick Query: *Have you seen a cuckoo lately? Cuckoo clock that is. Most cuckoo clocks run by a weight on a chain, just like tower clocks and grandfather clocks.*

Fast Fact: The new technology of the first tower clocks seemed so remarkable that some people could not believe that the clocks really worked. Some believed that people were hiding in the towers and moving the hands of the clocks!

Chronometer

In 1761 John Harrison built the first ship board clock that kept accurate time at sea. It worked reliably despite the pitching seas and damp conditions.

This clock was later perfected and came to be called a chronometer.

A Pendulum Timer

Find out how the pendulum length affects the time of each swing and verify Galileo's discoveries for yourself!

Materials

string, floss or thread
masking tape
washers or other easily
   attached weights
scissors
ruler or paint stir stick
large book
flat desk
timing device with a sec-
   ond hand
measuring device
marker

Get Started

Discuss the principle of the pendulum and how it was discovered.
Group children into pairs and provide each group with the above materials.

What to Do

1 Cut a length of string to equal the distance from the floor to the desktop.

2 Tape one end of the string to one end of the ruler and tie the other to the washer.

3 Place the ruler on the desk so that about 4" (10 cm) extends beyond the desktop.

4 Place the heavy object on the other end of the ruler to hold it in place.

5 When the pendulum is hanging at rest, measure the length of the string from the end of the ruler to the washer. Divide that measurement in fourths and mark the distance 1/4 of the way from the ruler towards the washer.

6 Bring the weight to one side and prepare to release it on the count of three.

7 When the pendulum is released, have one partner begin timing and the other count the swings.

8 The "timer" will announce when 10 seconds has elapsed and the "counter" will stop timing.

9 Record the number or swings that took place in the 10-second interval.

10 Reposition the string, taping it to the ruler at the 1/4 mark.

11 Repeat the swinging-timing sequence and record the number of swings in the 10-second intervals.

12 Compare the two numbers. The number of swings should double with the 1/4 shorter pendulum–verifying Galileo's principle that the beat of each swing is affected by the length of the pendulum.

Try This

• Experiment with various weights and string lengths.

Get in Gear!

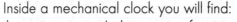

Inside a mechanical clock you will find:
the mainspring: it's the source of energy
the gear train: transmits the energy
the dial train: controls the movement of the hands
the winding and setting mechanism: lets you apply power and adjust the mechanism
the escapement and balance unit: controls the release of energy
the plates, also known as the framework: they enclose and protect the movement
jewels: provide bearing surfaces to reduce friction between tiny moving parts

Gears, or ratchet tooth wheels, inside a mechanical clock move at a steady pace, causing the hands to move at a rate consistent with the 24 hours of a day. Understanding how gears work will help you to better understand clock mechanisms.

Demonstrate:

Cut out cardboard gears so the teeth mesh together. Use paper fasteners to mount the gears to a base. Demonstrate how one gear turns another. Draw children's attention to the speed and the direction of the various gears.

Decorate a bulletin board with handmade gears and those found inside clocks or toys.

What Makes a Clock Tick?

2-4 • Science/Design and Technology/Art • Active Problem Solving/Exploration/Manipulation of Tools • 60 min

Caution: Extra supervision required.

The first inventors to make complicated mechanisms were the clock makers. Visit the Take-a-Part Center to find out about some of the mechanisms involved in clocks today.

Materials

old clocks and watches
screwdrivers
vise grips

Get Started

Collect discarded watches, clocks and other time-keeping devices for your collection of items. Review rules for safe use of tools and sharp instruments. Arrange for children to be supervised (but not guided!) at this center. Set up a take-apart center using these items.

What to Do

1 Inspire children to think about what makes the clock tick. What's inside?

2 Allow children to work at the Take-a-Part Center using tools to help them take a look inside.

3 Have children record some of their observations and findings.

4 Discuss their findings and their theories about how the clocks work.

5 Follow up with a look at the mechanics that help particular clocks keep time.

66

Gravity-Free Mechanical Clocks

In the 1730s an English clock maker named John Harrison built a clock that was entirely independent of gravity, with moving parts that were counterbalanced and controlled by springs. He had invented a new kind of mechanism to control the clock and compensate for outside influences. Although the pendulum clock was relied upon for exact measurements until the 1930s, it was replaced in popularity by new clocks that ran with gears, tightly coiled springs and manual winding.

In 1765 Harrison used his "clock knowledge" to make the first accurate chronometer - a clock that remains reliable in all weather, on land and at sea.

Quartz Crystal Clocks

In the 1920s scientists discovered that the vibration of a quartz crystal based on the action of molecules could keep time to within $2/1000$ of a second each year! By the 1930s this discovery was put to use in digital clocks and watches. The beat of these clocks is controlled by the vibration of a quartz crystal, not springs or gears. An alternation electric current supplied by an outlet or a battery caused the tiny quartz crystal to vibrate and display time as a digital display. By the 1960s this device was more accurate than the tuning fork timekeepers.

Fast Fact: Mechanical clocks and watches are being replaced by electronic and electrical timekeepers which are more accurate and easier to produce.

The Atomic Clock

Modern atomic clocks have now been developed in which the beat of the clock depends on the vibration of atoms.

In 1956 the atomic clock replaced the quartz crystal clock as the most accurate means of time measurement. The beat of the atomic clock depends on the vibration of electrons in atoms. These clocks measure microseconds and nanoseconds, keeping time to an accuracy of $3/100,000$ of a second - that's accurate to within one second over millions of years! The most accurate atomic clock in use is the cesium-beam clock. Atomic clocks provide accuracy that is needed by some scientists but not by most clock watchers.

Fast Fact: Many nations maintain precision cesium clocks from which times are averaged together to provide international atomic time (IAT). Highly accurate time signals from the world's national-standards laboratories are broadcast around the globe.

Pulsar Time

Scientists are always searching for more accurate time keeping devices. The most recent scientific discovery in timekeeping was the pulsar star. The pulsar star, discovered in 1967, may be the timekeeper of the future. The pulsar star emits many regular pulses of radiation per second - some keeping more accurate time than atomic clocks, but complex measurements of these pulses make this a difficult method of keeping time.

Make Your Own Time Measurement Contraptions

3-4 • Design and Technology/Science/Math • Problem Solving/Concept Reinforcement/Observation • 60 min

Materials

Center 1: beads, washers, strings, doweling

Center 2: marbles, tubes, gravity activated marble mazes

Center 3: water, waterwheels, water hoses or tubing, funnels, bottles, clock with a second hand 2 stopwatches

Get Started

Set up centers for each set of materials listed above. Display the materials for students to explore.

Divide the children into groups of two to six.

List three activities on the board or chart paper that could be timed: jumping up and down five times, 10 seconds (as indicated by the second hand on the clock) or clapping your hands five times.

What to Do

1 Talk about the ways that we measure time: seconds, minutes, hours, etc. Help children realize that this is a human-devised method for marking the passage of time. Watch the second hand on the classroom clock.

2 Ask children to think about ways that they can make up their own kind of timekeeping. How can they measure time? Demonstrate that counting can vary greatly in duration from person to person. Have one child count to 10 while another times her with a stopwatch. Have two children start timing on stopwatches and stop timing at the same time. The times should be the same when the watches are stopped. What was different between two children counting and two children using the watches? Help children to realize that the method of timing must be steady and consistent.

3 Explain that children will be allowed to create their own time-keeping devices using materials at their centers to explain how long the events on the board take to be accomplished. The time-keeping method must have something that moves at a steady interval.

4 Direct students to their centers where they will explore the materials and create a device that will time events listed on the board or chart paper.

5 Have each group demonstrate their device to the class. Help children to understand why some devices are acceptable timekeepers and others are not.

Fast Fact: Today's time-keeping mechanisms are simple to use and as precise as we need. Our most precise clocks measure time to a fraction of a second and are so accurate that they are in fact more consistent than the Earth's rotation itself.

Watches

As people began depending on time more and more, they needed a timepiece they could carry with them. Early clocks were too large and needed specific conditions to work properly.

The first clock small enough to be portable was invented in the early 1500s, in Nuremberg, Germany by, locksmith, Peter Heinlein. An hour hand was powered by the energy of an unwinding coiled metal spring. The device was not accurate enough to measure minutes and needs to be adjusted regularly. By the 1580s these first commercial watches were being produced in small quantities–only they weren't called watches–they were called Nuremberg eggs because they were about the size and shape of a goose egg.

In 1665 Robert Hooke invented the coiled spring to transmit power as it unwound. The vibrating spring controlled oscillations of the balance unit. This hairspring was the key to more accurate timekeeping and was used in all future mechanical watches. These timekeepers replaced hourglasses and provided consistent timing of duty watches aboard ships. Minute hands, glass protective covers and then second hands were added. In about 1700, jewels were added to reduce friction in the mechanisms. In 1753, clock maker Harrison invented a pocket watch that was as accurate as his clocks. In 1770 automatic winding was invented, and the watch was on its way to great popularity. The portable timekeepers became very fashionable and were worn as much for style as for practicality.

Quick Query: Why do you think that these early watches were called "pocket watches"?

Time on Your Wrist

Quick Query: About 100 years ago a time-keeping device replaced the pocket watch. Do you know what that device was? You may be wearing one right now!

It took until about 1880 for the idea of wristwatches to catch on. The Imperial German Navy recognized the practicality of this watch and had officers wear wristwatches aboard ships. People quickly recognized the convenience of wristwatches.

Quick Query: Would you prefer a wristwatch or pocket watch? Why?

Tuning-Fork Power

In 1953 the first commercially viable electronic watch hit the market. The mechanical mainspring of this device was replaced with a battery-powered tuning fork. It was accurate and affordable and brought watches to the wrists of many people.

Name _____

Why Do We Use Clocks and Watches?

Clocks and watches keep precise time all day and night. They make it easy for us to keep track of the _____ and _____ and sometimes _____. With only a glance at a clock or watch, we can know what time of day it is–once we have learned to tell time.

What would the world be like if we didn't have clocks and watches?

_____

_____

_____

_____

As life became more complicated, people needed to measure time in smaller and smaller units. Time-keeping devices were invented to keep the world operating smoothly and efficiently. Today, almost everyone uses a clock or a watch–they're everywhere!

Where do you see clocks? _____

_____

List clocks and watches that are used for specific time measurement tasks.

_____

_____

List five different kinds of clocks.

1. _____ 2. _____ 3. _____

4. _____ 5. _____

Why Do We Need to Know the Time . . . All of the Time?

Today it seems, every second counts. From the time we wake up until we go to bed, we need to know the time. We need to know the time for buses, trains, planes and ships, school, classes, work, lessons and appointments. We seem to live by the clock!

What things in your life depend upon specific times?

_____

_____

Tally the Timekeepers

Look around your home, school and neighborhood. Prepare a tally for adults and one for children.
Record the number of people who wear watches and the number who do not.
Make a pictogram to record the results of your study.
Is there a difference between the two tallies? Why or why not?
Do more children or adults wear watches?
What kind of watches did you discover?

Fashion Fad

Even before watches were accurate enough to be practical, they were a fashion fad throughout Europe. They were shaped of gold and silver, decorated with jewels and engraving and highly decorated hanging on chains, necklaces, belts and canes in many shapes and sizes. In the 18th and 19th centuries as watches were made smaller and smaller–they were sometimes found as rings and impractical earrings!

In 1934 the first Mickey Mouse watch appeared. In 1952 the digital electronic watches with the four-digit read-out was introduced and watch fashion was on its way. Today you can find watches of just about every shape and color that tell not only the time but also the date the times around the world. Some watches act as stopwatches or alarm clocks and some glow or wind up. Today it seems that watches are worn as much for fashion as they are for practical timekeeping!

Watching the Market

Bring in ads showing watches. What fashions do you like? What ones do you dislike? Compare the costs of the various watches. Prepare math exercises based on the watch prices.

Fast Fact: A watch's value may be based on the numbers of jewels contained. Between 7 and 23 sapphires, rubies or diamonds can be found within a watch mechanism.

Roman Numerals

Some clock and watch faces have numbers, some have none and some have symbols or just the numbers 1, 3, 6, 9 and 12. Some faces have Roman numerals, a set of numbers used long ago in ancient Rome. Use the chart below to learn Roman numerals you might find on a clock. (The numeral 4 is usually written as IV, but it sometimes appears on clock faces as IIII.)

Roman Numerals

1	2	3	4	5	6	7	8	9	10	11	12
I	II	III	IV or IIII	V	VI	VII	VIII	IX	X	XI	XII

Make Your Own Designer Watch

There are many styles of watches. Some are bold and bright, some are dainty and look more like fancy jewelry than functional watches. Some are waterproof, some glow in the dark and some tell you much more than just the time.

Design your own wristwatches. Add features you might need in a watch and then decorate to your own taste, with pens, markers, stickers or paint. Cut each watch out and then tape the watch band together around your wrist so you can show off your designer piece!

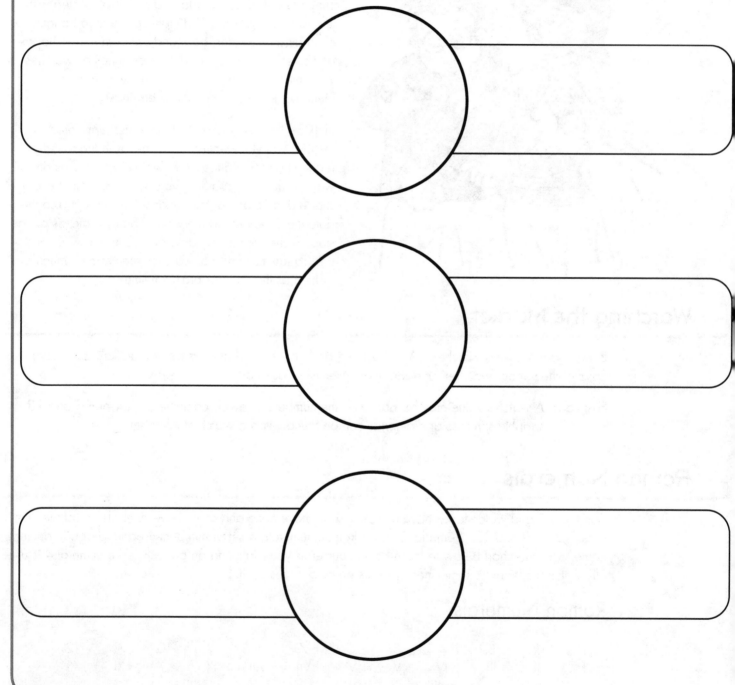

Chapter 6

Telling Time Using Clocks

Telling Time

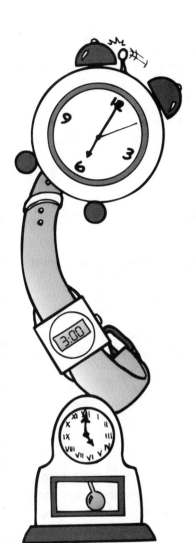

It's time to learn how to tell time.

Don't be shy you'll do just fine.

Sixty seconds will tick away

To become the minutes that make up our day.

Sixty minutes make an hour.

Learn this fact and you've got power!

There are 24 hours in each day,

Giving us time to sleep and play.

The clock will help us to tell the time,

If it's 3:00 or half past nine.

Clock faces and digital displays

Will help us all on our way.

We'll see numbers and hands, both short and long.

Follow the rules and you can't go wrong.

We'll unravel the mysteries of the clock

And have time left over to take a walk.

R. Eagan

Practice Make Perfect

Encourage children to use their newly found time-telling skills as often as they can. They can practice telling time anytime. There are clocks and watches everywhere.

Time Sweet Time

Introduce or conclude the time telling unit with a timely round cake. Bake and frost any round two-layer cake. Decorate it to look like an analog clock. Use a cake decorating kit to make icing numbers or purchase cake decorating numbers for your "clock." Position the numbers around the cake. Place a round candy in the center and use licorice twists for the hands. Use your creativity and candy or sprinkles to decorate the sides and make a fancy clock.

Fast Fact: About 5,000 years ago the Babylonians invented the system we still use to divide the day into 24 hours, with each hour containing 60 minutes and each minute containing 60 seconds.

Analog Clocks

The analog clock tells the time and provides a visual aid to reinforce the concepts of timekeeping. There is an hour hand, a minute hand and sometimes a second hand. The hands move around the clock face, in a clockwise direction, at a steady pace to indicate the passing of minutes and hours and seconds. The position of these hands tells us the time.

Clockwise Movements

Have students watch the second hand on the clock. Have them move one arm in a clockwise direction. Have they heard this term before? Now form a circle around the outside of your classroom and move in a clockwise direction to reinforce this idea. What is movement in the opposite direction called? Counterclockwise, of course.

When children have mastered an understanding of these directions, add some music and choose a caller. One child will announce "clockwise" or "counterclockwise" and start the music. Children move in that direction until another is called.

The Language of the Clock

There is a special language of the clock that helps people to communicate the times. There are a few different ways to tell the same time as you will see.

The Clock Face

Clock face refers to the clock surface that we see on our time-keeping device. Early clocks were numbered from one to 24 and had only one hand that moved around the clock face once. We still measure 24 hours in a day, but traditional clocks are numbered from one to 12.

Let's take a look at the face of the clock. What do you see? Children will probably notice the hands, the numbers and the little black markers around the outside of the clock face.

The Minute Markers

Did you notice the minute markers on the clock? Minutes are marked on most clocks. Look for the little black marks. How many of them are there all the way around the clock face? Sixty, that's right, one for each minute. Count the markers between each number. There are five, which means there are five minutes between numbers.

The Minute Hand

The minute hand is also known as the long hand. Can you find it on your clock? This hand travels around the clock face 24 times each day-once for each hour.

The Hour Hand

The hour hand is sometimes referred to as the short hand. Find the hour hand on your clock. In a day the short hand or hour hand travels completely around the clock twice-once for the 12 morning hours and once for the 12 night hours.

The Second Hand

There is a third hand on some clocks that's the second hand. This hand is usually very thin and moves very quickly around the clock face, going all the way around the clock once every 60 seconds, or once every minute. Do the clocks around you have second hands?

Quick Query: *How many times does the second hand travel around the clock in one minute? In one hour? In one day? Why might this hand be needed?*

Count Around the Clock

Get to know the numbers from 1 to 12. Cut out and color the clock face, trace the numbers, count around the clock and point to the numbers requested by friends.

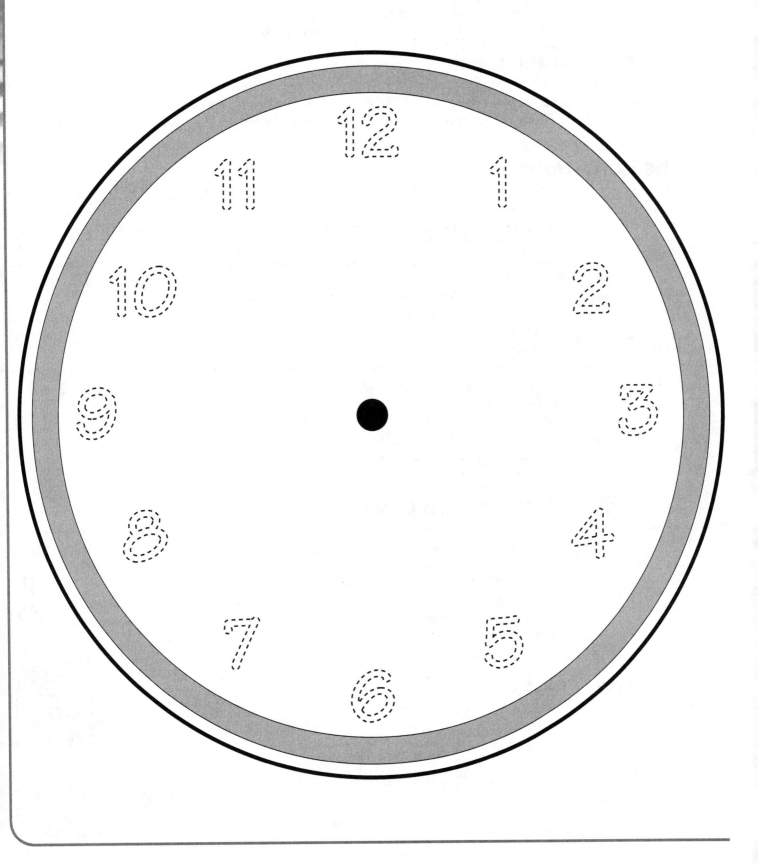

Let's Make Our Own Clock Instructions

Minutes on the Clock

How many black marks are there around the face of your clock? What do you think these represent? There are 60 minutes in one hour so there are 60 minutes marked a clock face.

Numbers on the Clock

The clock is numbered from one to 12. *Put the appropriate numbers in the correct circles on your clock. How many minutes are there between the numbers?* The numbers on the clock designate five-minute intervals of time.

The Long Hand

The long hand on the clock face is called the minute hand. It tells you how many minutes have passed. It takes 60 minutes or one hour for the long hand to go all around the clock face. This hand shows us how many minutes have gone by since the hour. When the long hand points straight up at the 12, it tells you that time is something o'clock. When the hand points to one, five minutes have passed. When it points to two, 10 minutes have elapsed. When this hand points to 11, it has travelled for 55 minutes since the new hour, and when it is back on the 12, 60 minutes have elapsed since the previous hour and the new hour begins.
Cut out the long hand on your sheet.

The Short Hand

The short hand is called the hour hand. It tells you the hour. It takes 12 hours for the short hand to go all the way around the clock face.
Cut out the short hand. Pierce the black dots on the short and long hands with a paper fastener. Then push the fastener through the dot on the clock to fasten. The hands to the clock face.
The short and long hands work together to indicate the correct time.

Tell Time with Your Own Clock

Use your clock to help you learn to tell time before moving on to the worksheets.
- *Point to the minute marks on your clock. How many are there in all? That means there are 60 minutes marked on the clock.*
- *Point to the minute hand or long hand. Move it around the clock one time counting the minutes by ones or fives as you go. When the big hand does this on a clock, it shows us that 60 minutes have passed. Sixty minutes is equal to one hour so this tells us that one hour has passed.*

Body Clock

1-3 • Active Game • Reinforcement • Short

Turn your body into a clock!

Hold a pencil in your right hand. That's your long hand–the minute hand. Your left hand will be the short hand–the hour hand.

One person calls out "o'clock" times, and the group makes their body clocks tell that time.

Let's Make Our Own Clock

Cut out the clock face and hands below. Fill in the numbers in the correct spots.

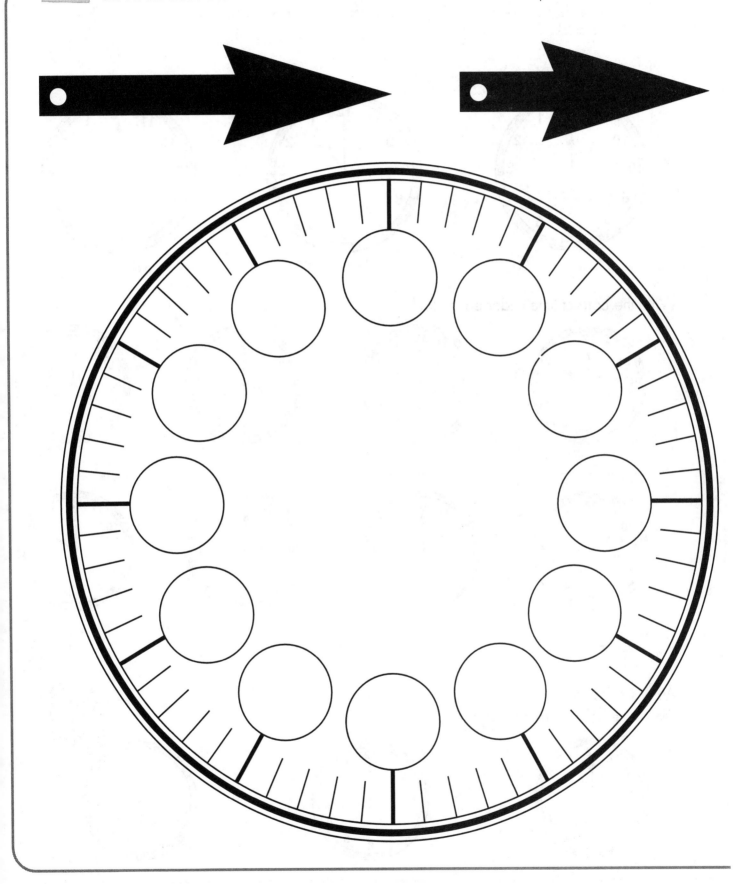

O'Clock

The long hand shows how many minutes have passed. When the long hand is pointing straight up at the 12, it means that a new hour is beginning and no minutes have passed that hour. We say that it is something o'clock.

The small hand tells us the hour by pointing at a number.

12 o'clock or 12:00   1 o'clock or 1:00   6 o'clock or 6:00

Write the correct time under each clock.

1. _____   2. _____   3. _____   4. _____

5. _____   6. _____   7. _____   8. _____

What's Missing?

Filling in the missing items on each of the clock faces.

1.    2.    3.

Fold an Hour in Halves and Quarters

*Have you heard people speak "time language" before? Have you heard the terms **o'clock, half past, something** or **a quarter to** or **after?** This can sound like a foreign language to someone who hasn't learned how to tell time!*

It takes the minute hand one hour, or 60 minutes, to go all the way around the clock face. The hand first travels one quarter of the way around, then one half, then three quarters of the way-breaking the hour into halves and quarters which are useful units when we talk about time inbetween the hours.

Fold an Hour in Half

Recopy the clock on page 77 or remove the hands and do this exercise. This circular clock face represents on whole hour-which is equal to 60 minutes.

Fold the left side of your clock over the right side of your paper-folding your clock in half. You have just folded an hour in half. You will see that the six marks the halfway fold. How many minutes are on each half of the clock? You can see that when the long hand reaches the six, it has travelled halfway around the clock, marking 30 minutes of time which is half an hour!

Half Past

If 60 minutes is a whole hour, then 30 minutes is half of that hour. When the long hand is on the 6, it has traveled 30 minutes, halfway around the clock face and we say it's half past the hour. (The short hand tells us what hour that is.) When the long hand is halfway around the clock, pointing to the six, we say it is half past the hour or something 30.

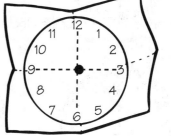

Fold an Hour into Quarters

Fold your clock in half as above and then fold it in half again, bringing the top of your paper to meet the bottom of your paper. Now unfold it. Your paper will be divided into quarters, or fourths, and so will your hour! How many minutes are contained in each section? Fifteen minutes is one quarter of an hour. How many quarters are shown on your folded clock?

Quarter After/Quarter Past

In an hour is 60 minutes, then 15 minutes is equal to one quarter of that hour. When the long minute hand is between the 12 and the 6 (at three), it has gone a quarter of the way around the clock face and we say that the time is quarter past the hour.

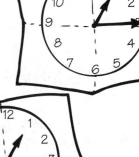

Quarter to

When the long hand is between the 6 and the 12, pointing to the 9, 45 minutes of that hour have elapsed, and it is 15 minutes until the next hour. The hand has gone three quarters of the way around the clock face and has one quarter to go, so we say it is a quarter to the hour.

Half Past

Half past (or $^1/2$ past) means that the time is _____ minutes or one half hour past the hour. There are two ways to write these times.

half past twelve, 12:30 or twelve thirty   half past six, 6:30 or six thirty

half past or $^1/2$ past = _____ minutes after the hour.

Write the correct time under each clock.

1.   2.   3.

_____   _____   _____

Quarter After and Quarter To

Quarter After

$^1/4$ *after* or *quarter after* means the time is _____ minutes after the hour. These are two ways to write these times.

quarter after twelve or 12:15   quarter after three or 3:15   quarter after six or 6:15

Write the correct times under the clocks below.

1.   2.   3.

_____   _____   _____

Quarter To

Quarter to means that $^3/4$ of the hour has passed and there is one quarter of an hour or _____ minutes until the next hour. There are two ways to write this.

11:45 or a quarter to twelve   2:45 or quarter to three   5:45 or quarter to six

Write the correct time under each clock.

1.   2.   3.

_____   _____   _____

Minutes After

The numbers on the clock mark the five-minute intervals between the hours. We talk about time as minutes past or to the hour. We say that is it so many minutes after the hour up until 12:30. After 12:30 we talk about how many minutes after the hour or how many minutes before the next hour.

Before 12:30

12:05 or five after twelve

12:20 or twenty after twelve

12:30 or twelve thirty

Write the correct time under each clock.

1.

2.

3.

4.

_____ _____ _____ _____

After 12:30

12:30 or twelve thirty

12:45 or quarter to one

12:50 or ten to one

Write the correct time under each clock.

1.

2.

3.

4.

_____ _____ _____ _____

81

Face the Clock

Answer the following questions and turn the circle into a clock face.

1. A clock is divided into _____ hours and _____ minutes.

2. The short hand is called the _____ hand. Color it orange.

3. The long hand is called the _____ hand. Color it blue.

4. There are _____ minutes between each of the numbers.

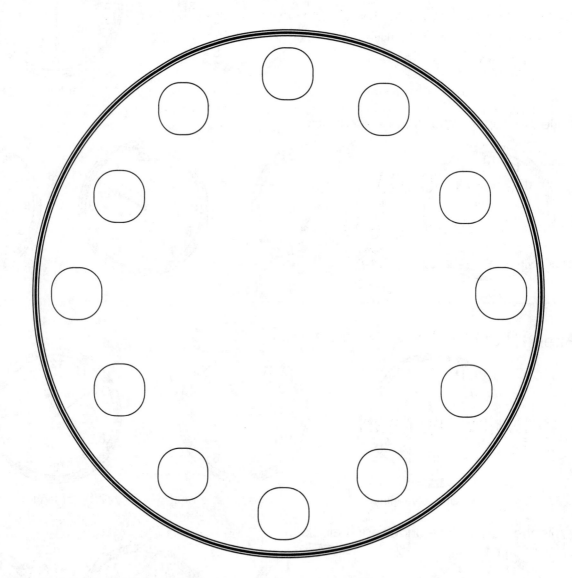

5. There are _____ minutes in an hour.

6. There are _____ hours in a day.

7. There are _____ seconds in a minute.

8. Each half hour equals _____ minutes.

9. Each quarter hour equals _____ minutes.

Digital Clocks

You will need a digital clock or a representation of one for this lesson.

The digital clock was invented after the analog clock. It doesn't have a face or hands to help us tell the time, but it does have numbers. The passing of time is shown on a digital display of dots and numbers.

Have you ever noticed a digital clock before? Where?

The digital display has four spaces for numbers. Two numbers go before two dots or a colon and two numbers go after. These two sets of numbers work together to tell us the time.

The first number in the spaces before the dots tells time like the short hand on an analog clock–these tell us the hour. These numbers go from 1 to 12. If the number displayed is 12, the time is 12 something.

Diagram: **12:00**

The number after the dots are like the long hand on an analog clock; they tell us the minutes. These numbers read from 1 to 60–the number of minutes in one hour. When 00 is displayed, it means that it is something o'clock and no minutes have passed since the hour.

Diagram: **4:00–or–04:00**

Quarter after the hour is displayed as: ___**:15**

Half past the hour is displayed as: ___**:30**

Quarter to the hour appears as: ___**:45**

When 60 minutes have elapsed, the numbers after the dots read 00 and the numbers before the dots change to display the new hour.

Digital Flip Chart

3-4 • Math/Language • Counting/Reading Time/Problem Solving • 10 min

Materials
4 binding rings
4 sets of numbers 0-9
1 colon symbol

What to Do

1 Hang the sets side by side with two sets of numbers on either side of the colon(:).

2 Change the times by flipping the numbers. Have children say the times that are displayed or make the chart read a time given by another child.

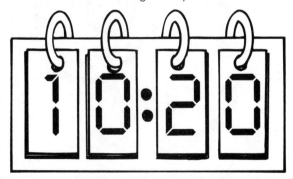

Digital Time

Write the correct time beneath each clock.

1. | 1:15 | 2. | 11:30 | 3. | 2:45 | 4. | 3:20 |

_____   _____   _____   _____

Fill in the clocks to read the time printed below each.

5. | : | 6. | : | 7. | : | 8. | : |

one o'clock   noon   three thirty   quarter after ten

Match the clocks to the time shown.

twelve noon   three thirty   six o'clock   half past six

9. | 6:30 | 10. | 6:00 | 11. | 12:00 | 12. | 3:30 |

Draw a line between the matching digital and analog clocks.

 | 12:00 |

| 7:15 |

 | 4:45 |

| 2:30 |

Digital and Analog Time

What Time Is It?
Write the correct time under each clock.

1.

_____

2.

_____

3.

_____

4.

_____

The Time Is . . .
Draw the hands or write the digital time on each clock to make it read the correct time.

5.

1:00 a.m.

6.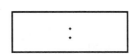

ten past three

7.

4:00 a.m.

8.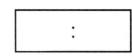

twelve o'clock

Write the digital time beside the analog time shown.

9.

10.

11.

12.

13.

14.

Name ____________________

Clock Match

Find the matching pairs of times. Cover the matching pairs with coins or tokens or cross them out. What time is left?

1:00

11:00

3:20

6:15

4:55

2:30

9:45

10:50

2:30

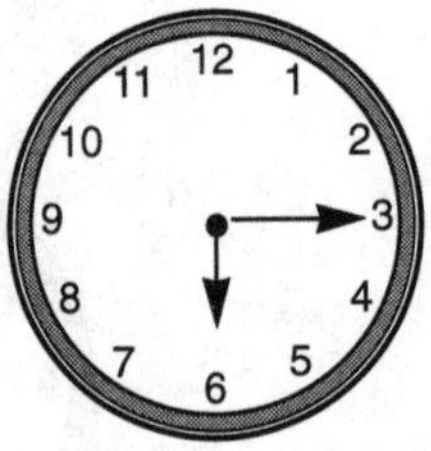

4:10

A Test of Time

1. The clock is divided into _____ hours.

2. There are _____ minutes in each hour.

3. One half hour equals _____ minutes.

4. One quarter hour equals _____ minutes.

Fill in the face on the clock.

Write the time below each clock.

5.

_____

6.

_____

7.

_____

8.

_____

Add hands to make the following clocks read the time written below.

9.

1:15

10.

7:20

11.

ten past eight

12.

2:15

Write the digital time for the following clocks.

13.

14.

15.

16.

The Clock Strikes 12

An "timely" educational card game

Materials

1 deck of cards with the Jokers taken out

Object

To be the first player to place all of their cards in clock formation.

What to Do

1 Shuffle the deck and place it facedown in the center of the table.

2 Familiarize players with the values of each card. In this game the Ace represents one, 2-10 represent their values. The Jack is 11, the Queen is 12 and the King is a wild card and can be any number.

3 The first player chooses the top card off the deck and turns it over. The player then places that card in the correct position to make her clock.

4 The next player turns over a card and places it as the first player did. When a player draws a card that has already been placed on her clock, that card is placed facedown beside the drawing pile. If the drawing pile is depleted, players draw from the discard pile.

5 Play continues in this manner until one player completes her clock.

Set Your Clocks

Materials

1 set of time cards
1 clock face (page 77) for each player

Get Started

Make a set of cards to show digital or spoken time.

Choose one player to be timekeeper. Have each player make the clock on page 77 or supply ready-made clock faces.

Object

To set your clock face to the correct time.

What to Do

1 The timekeeper pulls one card from the deck and reads the time to all players.

2 Each player sets their clock to the time stated.

3 When all players have set their clock to the correct time, a new time card is selected.

4 Continue to play for a specified time period or for a pre-determined number of card selections.

Try This

• Give points to the first player who sets their clock correctly.

What Time Is It, Mr. Wolf?

K-3 • Physical Education/Math • Active Movement/Counting/Time Language/Listening • 10 min

Materials

chalk (may not be necessary)

Get Started

Find an area with two boundaries clearly marked from 20-50 feet apart–chalk lines on pavement, the end of pavement, gymnasium boundaries, a fence line or wall will work. Designate the area behind one boundary as the Wolf Zone and the area behind the other as the Safe Zone.

Choose one child to be Mr. Wolf.

Object of the Game

To be the first child to reach the Wolf Zone or be the only player not tagged by Mr. Wolf to become Mr. Wolf yourself.

How to Play

1 Mr. Wolf stands on the boundary that designates the Wolf Zone. He faces away from the group.

2 The other players stand on the boundary that designates the Safe Zone and call out, "What time is it, Mr. Wolf?"

3 Mr. Wolf answers with a time, i.e. "One o'clock."

4 Players take the number of steps designated by the hour given–always moving closer and hoping to cross Mr. Wolf's boundary line and beat Mr. Wolf at his own game. If this happens, which it rarely does, the player who crosses the boundary becomes the new Mr. Wolf and the game begins again.

5 At any point, Mr. Wolf can call out "Dinnertime," and turn to chase players. When players hear "Dinnertime" they run back to the boundary line where they began and cross into the Safe Zone where they cannot be tagged. Any player that is tagged before reaching the Safe Zone becomes a wolf and joins Mr. Wolf in the next round of time calling and dinner chasing!

6 Eventually there is only one child left. This cunning, speedy player becomes the new Mr. Wolf for the next round of this game.

Strategy

Mr. Wolf may call out times that bring player close to him before he calls out, "Dinnertime."

Variation

Miss, Mrs. or Ms. Wolf will work for this game, too!

The 24-Hour Time System

An ordinary clock doesn't tell us if a particular hour is before noon or after noon.

Sometimes it is simpler to number the hours of the day from 0 to 24, eliminating the need for a.m. and p.m. The system of telling time that does this is called the 24-hour system. It is used in some countries to tell the time and in most countries for military, sea and transportation purposes. Look at an airplane or train ticket. Some countries say the time according to the 24-hour system even though they use the 12-hour clock face.

The 24-hour clock face marks all 24 hours–the divisions are very small and can be difficult to read. The hour hand makes only one revolution every 24 hours. Some digital clocks clearly display the hours and minutes of the 24-hour clock time.

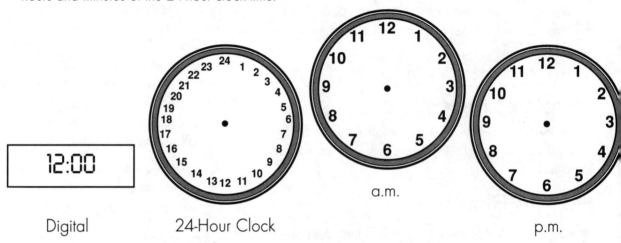

Digital   24-Hour Clock   a.m.   p.m.

In this system the beginning of the day is midnight or zero hour. The hours are numbered from 0 to 24 and written in four digits as 00:00 to 23:59. The first 12 hours are counted from midnight to noon and after noon; they are counted from 13 to 24. Anytime before 12 is a morning hour and anytime greater than 12 is an after noon time.

These times are referred to in "hundreds" and converted like this:
1:00 a.m. is written 01:00 and referred to as "Oh one hundred hours."
Noon is written 12:00 and referred to as "twelve hundred hours."
Minutes are written in the usual way from _ _:00 to _ _:59.
2:36 a.m. is written as 02:36 and referred to as "Oh two hundred thirty-six hours."
1:01 p.m. is written as 13:01 and referred to as "Thirteen hundred, oh one hours."

To convert standard time to 24-hour time, put a zero in front of any hour or minute with a single digit in the a.m. hours, and add 12 to the hours in the p.m. times.

To convert from 24-hour time to standard you can subtract 12 from anytime after the twelve hundred hour times to get the p.m. hour and remove any 0 that precedes the hours for the a.m. times. Digital times in the 12 hundreds or earlier are easily converted by ignoring any preceding zeros and reading the time as it is.

Quick Query: *Using this system, what time do you wake up, go to school, eat supper and go to bed?*

It is bad luck to have more than one clock in a room.

Chapter 7

Calendar Time

Try to imagine life without a calendar. What would it be like? Could we get by today without a calendar? How does it affect your life?

What Is a Calendar?

Have you seen a calendar? What does it look like? Have you seen one? Take a look at the calendar in your classroom or house.

The calendar is like a clock that helps us to keep track of the days of the year and the years. This system allows us to order the days and months, to determine when the year begins and ends and to keep track of the cycles of the sun, moon and seasons. It is a tool we use to count, record and organize the passing and the future days.

Calendars help to keep our society–and our lives–organized. They tell us the day of the week a particular date will fall on, the moon phase to expect at a certain time, remind us of upcoming events, allow us to record events in our lives and let us know when certain events like birthdays and most holidays will occur.

Our calendar contains 12 months of 28-31 days equalling 365 days a year–with the exception of the leap year which includes February 29 and equals 366 days. The days of each month are shown on one block or page.

Where Did Our Calendar Come From?

Our calendar arose from ancient calendars developed by early peoples' awareness of the cycles of the moon, the Earth and the sun to indicate the time. Their survival depended on the calendar which ordered planting, harvest and agricultural ceremonies; festivals and worship. Days were arranged into an orderly plan that counted and recorded the days into calendars of scratch marks in the dirt, carvings on stone pillars, writings on paper and even complex mechanical models and monuments such as Stonehenge which indicated the position of the sun and the time of year.

It seems that people have always understood that time passes, but developing a system to accurately measure time has been a challenge. From the times of the earliest Sumerian calendar to the modern Gregorian calendar, humans have been searching for a way to align the natural cycles of the moon, the Earth, the sun and the stars with a calculated system of time measurement.

Fast Fact: The word *calendar* derived from the Greek word *kalend* meaning "I shout" and the Roman word *calends* which was the first day of the Roman month.

Charting the Natural Cycles

The days, months and years that make up our calendar are based on natural rhythms: the 24-hour day/night cycle: the monthly waxing and waning of the moon; the four seasons and the Earth's annual trip around the sun. The days, months and years are based on natural cycles.

The Solar Year

Our Earth spins on its axis and moves in an orbit, with the moon, around the sun. Our year is based on the time it takes for the Earth to travel in its circular orbit once around the sun. This journey provides a reliable constant by which they could measure a year. The complete cycle takes 365.26 days–the unit of time we call one solar year. Sun-watchers figured out that a year was the number of days it took for the sun to move, or so it seemed, from one position in the heavens back to the same place again.

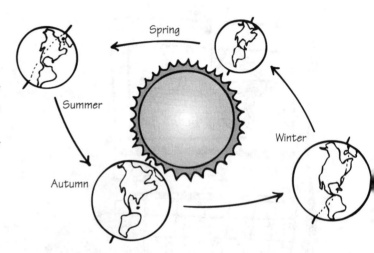

One thousand and five hundred years ago, the Mayans of Central America accurately calculated the length of the solar year and created a calendar that was as accurate as the one we use today!

Fast Fact: The Latin word *sol* means "sun." Can you think of any words that include this Latin word?

Seasons

The year is divided into four quarters. Each begins with a special sun day. On the spring and autumn equinoxes (the first days of spring and fall) we experience 12 equal hours of daylight and darkness. The winter and summer solstices are the first days of winter and summer. In the winter there are more hours of darkness and in the summer more hours of daylight–the amount of difference varies depending upon where you are.

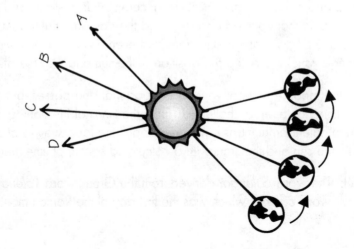

The Lunar Time: Months

The moon provided another predictable measure for early timekeepers – in fact it still does! The moon changes from waxing (growing bigger) – the barely visible new moon – to a slim crescent, a half-moon, an oval, or gibbous to a full moon and then wanes (grows smaller) back to a gibbous, half-moon, crescent and obscure new moon again. The phases of the moon always span 29 1/2 days – the length of time it takes the moon to travel around the Earth – known as the lunar month.

The time between one new moon and the next is called a lunar month or a month. As the moon orbits Earth it doesn't really change shape, it just appears to. Different parts of one side of the moon reflect sunlight making that part visible from Earth.

Our current calendar months are roughly based on the lunar month. The months no longer begin with a crescent moon as they once did. The phases of the moon can occur at anytime during a month. Take a look at your calendar or up into the night sky to track the phases of the moon.

Fast Fact: There is really no such thing as moonlight. The light we see is sunlight reflected off the moon!

Fast Fact: Native North Americans marked the time by moon months and referred to the past in terms of "many moons."

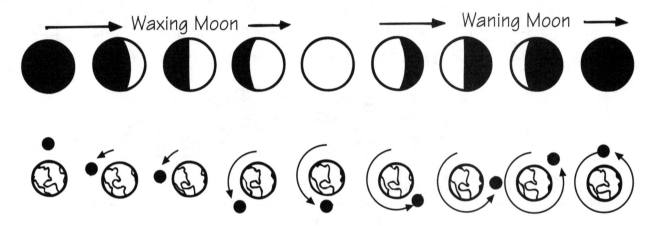

Moon Watch

Begin a Moon Watch when there is a new moon. Track the phases of the moon by observing and recording one date and moon shape in each block of the Calendar copy page (page 102). Check the sky every night. If it's overcast, check the next night or the next! Binoculars or a telescope can be used to enhance the moon watching and bring lunar features into view. Take this opportunity to introduce children to the "moon report" in the newspaper.

Fast Fact: Evidence has shown that ancient civilizations were aware of and recorded the moon phases as long as 30,000 years ago.

Keep the Calendar Months Straight

Look at your calendar; count the months, days and read the names of the months.

Quick Query: *Do all months begin on a Sunday? Why not? Do all months have the same number of days? What days occur on the same date each year?*

There are 12 months which make it easy to arrange the days of the year. Calendar months are divided into a cycle of seven-day weeks with about four weeks in each month. Some months have 30 days, some have 31 and February has 28, except for every four years when it has 29 days! How can kids (and adults) keep it all straight? Try the following memory aids to increase your calendar power!

The Months of the Year
Thirty days hath September
April, June and November.
All the rest have thirty-one
Save February
Which alone has twenty-eight,
Except for Leap Year–that's the time
When February's days are twenty-nine.

The Knuckle Calendar

This old memory aid will help to teach the sequence and the lengths of the months.

Form one hand into a fist. Using the finger of the opposite hand, point to the first raised knuckle and then the depression between the knuckles, the next knuckle and so on. When you reach the last knuckle, go back to the first knuckle. Name one month as you point to each knuckle or depression. Months that are named when you point to a knuckle are the months with the higher number of days; the months that are named when you point to the depressions are the months with the lower number of days.

Quick Query: *Which month has 28 days? (All of the months do!)*

Fast Fact: According to folklore, you will have good luck if you turn the silver in your pocket when you see a new moon.

Combining the Lunar Month and the Solar Year

Our year is based on the sun, but our months are based on the moon. Trying to combine the two has caused calendar problems throughout time. Early people recognized that the 12 lunar months were almost equal to one solar year. The lunar year (based on twelve 29 $1/2$ day moon months) is $11 1/4$ days shorter than the solar year. From the earliest days, humans attempted to coordinate the lunar months. with the solar year while keeping the natural cycles in step!

Calendars often ended up so out of sync with the natural cycles that they were no longer useful to the people who needed them. An accurate means of time measurement was needed and eventually developed.

Moon Facts: *Luna* is the Latin word for *moon*. The Jewish calendar is based on lunar months and solar years with adjustments made along the way. The Islamic calendar is based on the lunar calendar, and the Hindu calendar is based on the solar year with religious holidays established by the lunar cycle.

The Leap Year

The leap year doesn't occur to mix us up–it occurs to keep the lunar calendar in sync with the solar calendar. It takes Earth 365 days and six hours to orbit the sun. After four years those extra six hours add up to an extra day in the year. That day becomes February 29, and that year will have 366 days. That's called a leap year.

Leap Through the Years

Leap years occur in years that can be divided equally by four–except for the centurion years which must be divided equally by 400 to be leap years.

Write some leap years on the board and leave some blanks. Can students determine the pattern and fill in the blanks?
Try these sequences:

    1904, 1908, 1912, _____, 1920, _____, 1928, 1932, _____
    or
    1880, 1796, 1792, _____, 1784, _____, 1776, _____

Quick Query: *Will the year 2000 be a leap year?*

Fast Fact: During the French Revolution at the end of the 18th century, the base 10 metric system was introduced for timekeeping. The month, year and 10-hour day were not consistent with natural rhythms and soon disappeared from usage.

Ancient Calendars

The early calendars were developed by astronomers who studied the stars, planets and other heavenly bodies; astrologer who predicted events by the position of the heavenly bodies and by religious leaders. They provided guidance and instruction to their followers. In some cases it was priest-astronomer or astronomer-astrologers who helped to shape the early calendars – generally a combination of skills came together in the making of the calendars.

The early calendars helped to determine important dates for planting, harvesting, worshipping and celebrating. With the exception of the Mayan calendar, the early calendars arranged the days with a margin of error that grew and grew until the dates and the seasons were no longer connected. Some ancient calendars were abandoned and others were lost when civilizations were destroyed by war or new religions. Useful elements of many calendars, and the concepts they were based on, led to the calendar we use today.

Fast Fact: Calendars were created by the Aztec, Babylonians, Chinese, Egyptians, Romans, Hebrews, Mayans – well, you get the idea, almost every civilization had their own calendar, usually based on an understanding of the lunar month, the solar year or both.

Sumerian Calendar

The ancient Sumerians of Mesopotamia were good timekeepers. Five thousand years ago their priest astronomers developed a lunar calendar with months divided into seven-day weeks, days into 12 Sumerian hours and hours into 30 minute-like units.

The Celts

The Celts, or ancient Europeans, divided the year into only two seasons – summer (the sun's strength) and winter (the sun's weakness). Almost 4,000 years ago, in England, they built a most interesting calendar called Stonehenge. Stones weighing as much as 50 tons were moved from Wales to construct the 100-feet-in-diameter and 30-feet-high calendar. Two stone circles and half circles, an altar stone and a heel stone keep track of changing seasons. On the longest day of the year, from the altar stone, the sun appears to rise directly over the heel stone casting a shadow that reaches across the altar stone. On the longest evening of the year, from the same position the moon rises between the two inner-circle stones that frame the heel stone. This calendar marks the equinoxes of the two seasons.

Babylonian

In about 1750 B.C. the Babylonian's revised the Sumerian calendar to create a calendar 12-month, 354-day lunar calendar, with seven-day weeks based on the worship of the sun, moon and five know planets: Shamash (Sun/Sunday), Sin (Moon/Monday), Nebo (Mercury/Tuesday), Ishtar (Venus/Wednesday), Nergal (Mars/Thursday), Marduk (Jupiter/Friday) and Ninurta (Saturn/Saturday). The number seven was very important and magical – signifying the seven most prominent heavenly bodies. This seven-day cycle was adopted by the Jewish and Egyptian people, but the lunar calendar was eventually abandoned as it was 11 1/4 days short of the solar year.

TLC10073 Copyright © Teaching & Learning Company, Carthage, IL 62321-001

Egyptian

In ancient Egypt, life revolved around the river Nile. The flooding of the Nile began on or near the time when the bright star Sirisu reappeared on the horizon just before sunrise, marking the beginning of a new year. A 52-week calendar was developed based on Sirrius and the stars. The year was divided into 12 thirty-day months, with an extra five days tacked onto the end of the year to keep it almost in sync with the solar cycle. The 365-day year was short $1/4$ of a day per solar year and fell behind the natural seasons, one day every four years.

Mayan

Almost 2,000 years ago, the Mayan civilization of Central America had a very advanced understanding of time. They believed that the past and future were the same and that time had no beginning or end. Their religious beliefs and agricultural practices were based on the cosmology and relied on a combination of the solar year and their sacred round calendar. Priest-astronomers observed nature and the heavens to make precise records in almanacs and on tall stone pillars. Unlike other early civilizations, the Mayans were able to accurately calculate the length of the solar year to be just over 365 days. The calendar they developed was more accurate than the one we use today!

Native North American

Many Native people view the passage of time as circular and used a circular calendar instead of a linear one. The sacred medicine wheel shows the cycle of seasons and the cycle of life. It is lined up with the four directions–north, east, south and west. The Kwakiutl on the northwest coast of North America and The Mistassini Cree of Quebec divided the year into two seasons–summer and winter.

Roman

Although there have been many changes over 2,500 years, the calendar we use today comes from the early Roman (and to some extent the Greek) calendar of the 8th century B.C. The first Roman calendar was based on the $365 1/4$-day solar year but consisted of 10 months which added up to 304 days. The last month was extended for about 60 days until a new cycle began in the heavens. In 712 B.C. King Numbe added Januarius and Februarius to the year and changed the lengths of the month and in about 500 B.C. January and February were added.

The Julian Calendar

By 46 B.C., the Roman calendar was out of sync with the seasons by about 90 days. Julius Caesar made adjustments so the calendar would repeat itself annually and stay in close step with the solar year. That particular year had 445 days! The months had 30 or 31 days except for February with 28–bringing the day count to 365. Because the solar year was actually $365 1/4$ days, one extra day–February 29–was to be added every four years, introducing the leap year. His successor, Caesar Augustus made further alterations to this Julian, or Old Style, calendar and it remained in use for over 1,500 years.

The Roman Months

Martius, Aprilis, Maius and Junius: so called after gods and goddesses
Quintilis: meaning the fifth month, later renamed *Julius* in honor of Julius Caesar came to be called July
Sextilis: meaning "the sixth month" renamed *Augustus* in honor of Caesar
Augustus came to be called August
September: meaning "the seventh"
October: "the eighth"
November: "the ninth"
December: "the tenth"
Later Additions
January: named for the god Janus
February: an ancient purification feast

The Gregorian Calendar

You probably call the calendar you use today "the calendar." Did you know that it is also known as the Gregorian calendar, the New Style calendar, the Christian calendar, the common calendar *and* the standard calendar? This calendar-whatever name you choose to call it-underwent many revisions to become what it is today!

The Julian calendar was based on a 365¼-day year when the actual solar year was 365 days, 5 hours and 49 minutes or so. By 1582 the extra 11 minutes had become 10 days. In 1582 Pope Gregory XIII took 10 days from the calendar to put the Julian calendar back in step with the sun. To keep the calendar in step, Pope Gregory determined that the century years would have to be divisible by 400 to be a leap year. He also changed the beginning of the new year from the ancient date of March 25 to January 1. Many resisted the changes of the Gregorian calendar, but by 1752, Britain, which included Canada and the colony of America, had adopted it and by 1950 it was in use just about everywhere in the world.

Numbering the Years

The Romans began numbering the years to help people distinguish between the passing years. Year one was initially attributed to the year the Roman Empire was founded. This numbering system continued until the year 1200 when Christianity became the official religion of the Roman Empire. A Christian monk named Dionysius Exiguus proposed that year one should be the year of Jesus Christ's birth-which he established to have occurred 527 years earlier. Dionysius's date was accepted and the Roman Year 1200 was changed to 527 A.D. All years after 1 are referred to as A.D. from the Latin phrase *anno domini* meaning "the year of the Lord," and all years that came before are referred to as B.C., "Before Christ." The numbering of the years in this manner became widely used by Christians and non-Christians alike and is in general usage today with exceptions of the Jewish religious calendar dating from the creation of the world and the Islamic calendar dating from the flight of Mohammed to Medina.

The Search for the Perfect Calendar

The Gregorian calendar will be in step with the sky events and the seasons for thousands of years, but it is off by 26 seconds a year.

Some people find this calendar confusing and have tried to improve upon it. The most recent and most popular improved calendar, the World Calendar, was developed in 1930 by Elizabeth Achelis. It repeats itself, exactly, year after year. Each year contains a basic 364 days, 52 weeks, with extra days added at the end of December and Leap Year at the end of June. Every year begins on a Sunday, January 1, and every date, including holidays, will fall on the same day of the week-every year.

How would you feel about adopting this new calendar? How might it affect your life?

The Unnatural Week

You can figure out what time of the month it is by looking at the moon, what season it is by studying the weather and what time of year it is by mapping out the position of the sun and the Earth. But we cannot tell what day of the week it is by looking at the moon, the temperature or the sun on the horizon.

Although each phase of the moon lasts a little over a week, there is really no natural rhythm by which the week was formed–it is one of the "man-made" units of time measurement that can only be tracked by following a calendar. The week which is used throughout most of the world is related to the ancient Hebrew or Jewish calendar with the seven-day idea adopted from the Babylonians. In most countries, the week is now seven days long, with four to a month and 52 to a year. But it hasn't always been this way. In the past, weeks have been as short as four days and as long as 10!

Who Named the Days of the Week?

Ancient Babylonian stargazers were fascinated by the movement of what they called "the seven heavenly wanderers"–the sun, the moon and the five planets closest to the Earth. These wanderers were given the names of gods, and each god was given one day to rule.

At first, the weeks followed the phases of the moon. But each phase is about seven days long and the cycle could not begin again until the moon was full. There were only 28 days in each four-week cycle, but could there be as many as 31 days between full moons. There were always a few days left over at the end of every fourth week. To make matters worse, the importance of the gods–and thus the order of the seven days–varied from town to town!

Sometimes before the year 650 B.C., astrologers solved the problem once and for all–by naming each day according to the planet that "ruled" its first hour and by putting them in order. They didn't bother to wait for the full moon at the end of the month–they started the cycle over as soon as the fourth week was finished. The weeks have now followed one another, without a break, for more than 2,600 years!

Fast Fact: Even though the days of the week had been named by the time the Bible was written, the Old Testament refers to them only by number. This is because the seven days of the week were labeled by astrologers whose study was regarded as a strange and superstitious (if not evil) cult of star magic. In the Bible, the days of the week are referred to by number, a system still used in Russia, Greece, Portugal, China and the eastern European nations.

The Mithraists, a religious group that worshipped astrology, spread the days of the week westward from Babylon. Their influence was so strong that by the year 250 A.D., the week, as we know it, was used in England, France, western Germany, Austria, Italy, the Balkans, North Africa, Egypt, Palestine and Turkey. Eventually, the Teutons, found their own god and goddess equivalents for the Babylonian planet gods. In England, the Teutonic language (Anglo-Saxon) slowly developed into an early form of English, and the words for the days of the week have changed little since then.

Days of the Week: From B.C. to 20th Century A.D.

Babylonian Name	English Name of Planet	Teutonic Name of Day	English Name of Day
Shamash	Sun	Sunnandag	Sunday
Sin	Moon	Monandag	Monday
Nergal	Mars	Tiesdag	Tuesday
Nebo	Mercury	Wodnesdag	Wednesday
Marduk	Jupiter	Thorsdag	Thursday
Ishtar	Venus	Friadag	Friday
Ninurta	Saturn	Saeternesdag	Saturday

Fast Fact: A fortnight (a period of 14 days) is commonly referred to in England.

Is Sunday the First Day of the Week or the Seventh?

Although Monday was often referred to in ancient texts as the first day of the week, most people now believe that Sunday-"the day of the sun" (the most important "planet")-is the first day of the week and Saturday (named after Saturn, the unluckiest planet) is the last. In some countries, including China, Sunday is still referred to as the seventh day.

What About Weekends?

In ancient times, the first day of the week was celebrated, while the last day of the week was feared. This last day, related to our Saturday, was an unlucky day of rest, with many taboos. On this day, it was considered improper to engage in many activities including working, eating, correcting a child, burying the dead, making music, changing clothes, dressing in white, boasting, healing the sick, issuing royal decrees or making wishes!

Later, there were elaborate laws about working on the "Sabbath," which was set aside as a day of rest and worship. The rules were often bizarre and difficult to obey. Is it possible to carry nothing heavier than a fig? Why could you pluck one ear of wheat but not two? Most poor people had neither the time nor the desire to learn or practice all of the rules. For their defiance, they were expelled from the synagogues. As a result, they gradually stopped meeting on Saturdays to pray and concentrated instead on Sundays. Today, Jewish people regard Saturday as the Sabbath, while Christians worship and rest on Sunday.

Quick Query: *What is your favorite day of the week? Why?*

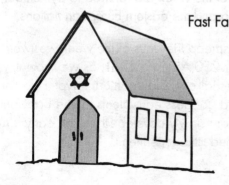

Fast Fact: Early Christian preachers believed that a day of rest was a good idea-even for slaves! This "good idea" became law, and some people were forced to pay fines if they did not attend church on Sunday. In England and North America, laws forbid people from playing games, making music, dancing, drinking alcohol, or buying or selling on Sundays. Even today, contracts are not always considered valid if they are signed or take effect on a Sunday, although "consumerism" has persuaded many businesses to stay open on Sundays, despite the penalties involved.

Year Board

This is a great activity to bring in the new year and brighten a classroom! The visual activity will lead children to a better understanding of the seasons and the passing of the months of the year.

Materials

magazines
photographs
old calendars
glue and glue sticks
scissors
3 pieces of bristol board or mural paper

Get Started

Prepare a large Year Board. Divide the board or paper into 12 sections.

What to Do

1 Explain that each section of the board represents one month. Review the names of the months as you write them in order in each section.

2 Have students draw, paint, write words or cut and paste magazine pictures, photographs or appropriate items in the area designated for each month.

3 Decorate a bulletin board or classroom wall with this visual calendar. Add letters to say "All in a Year." Hang calendars, pictures of the sun, the moon and phases of the moon, diaries, day timers, schedules, clocks, watches and other time-keeping devices around the border.

101

Make Your Own Calendar

Develop a better understanding of the calendar by making one of your own! Complete one page for each month to complete one group calendar. Label the month and week days, add the appropriate number of days for that month - number correctly and decorate the calendar. Mark special days with stickers, crayons, markers or pictures.

Saturday					
Friday					
Thursday					
Wednesday					
Tuesday					
Monday					
Sunday					

Holidays, Festivals and Important Dates

The calendar keeps track of important dates for us. We get a sense of the passage of a year with annual celebrations, festivals and holidays that are part of our calendar. The calendar tells us when it's a new year; our birthday; Mother's Day; Father's Day; Christmas; Hanukkah; Cinco de Mayo, an anniversary; the first day of school and other such religious, national and personal dates.

Many of the special days we celebrate come from ancient festivals and ceremonies that early people hoped would ensure their supply of food, water and good weather.

Quick Query: *What important dates help you to recognize the passing years?*

The Birthday Celebration

Birthdays are a great time to celebrate! You celebrate your first birthday after you have lived one year and are beginning your second year of life. In the year between birthdays, the Earth travels one complete path or orbit around the sun and returns to the position it was in when your were born.

Quick Query: *How many complete trips has the earth made around the sun since you were born?*

What Child Are You?

Many ancients believed your personality was determined by the day on which you were born. Recent week-wizards turn to this poem:

> Monday's child is fair of face.
> Tuesday's child is full of grace.
> Wednesday's child is full of woe.
> Thursday's child has far to go.
> Friday's child is loving and giving.
> Saturday's child works hard for a living.
> But the child that is born on the Sabbath day
> Is bonny and blithe and good and gay.

Find out what day of the week you came into the world. Does the poem hold true?

Animal Signs

Some Chinese people believe that the year you are born determines your animal sign-which affects your personality.

Planet People

Astrologers believe that the month you were born determines your zodiac sign-which is determined by the position of the heavenly bodies on your birthday. This sign is believed to affect your personality and day-to-day events in your life. Find out your sign. Do you fit the description?

The Birthday Challenge

Birthdays are wonderful celebrations that come once a year. They help us to keep track of how old we are.

When is your birthday? _____

How old are you? _____

What year were you born? _____

Match the birthday cakes to the correct children.

Decorate the birthday cake
and put candles on top to show
how old you are.

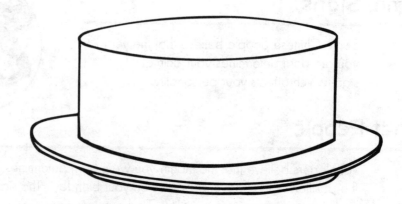

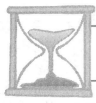

Chapter 8
Biotime

Most living things have a built-in sense of time. Human beings are born with a natural rhythm of sleeping and waking, and live their entire lives around the 24-hour spin of the Earth. So do animals, fish and even plants. But, unlike humans, who remain active year-round, many of these living things must also abide by a seasonal clock and plan for long periods of inactivity. These plans follow a regular annual cycle. Birds, fish and animals that cannot cope with extreme cold, for example, *migrate* (travel) long distances to winter in warmer climates. Some species, including tortoises and bears, *hibernate* (sleep) during the winter, slowing their normal life processes so much that they can survive for months without eating! Plants follow a seasonal cycle of growth, reproduction and dormancy or death.

The natural rhythms of plants, animals and people are named according to the Earth's motions, which tell all living things how to "set" their internal clocks.

Circadian Rhythm

Circadian rhythm is the 24-hour day/night cycle. It is named for the Latin words *circa*, meaning "about," and *dies*, meaning "day." All living things alternate their periods of activity and rest in circadian rhythm.

Tidal Rhythm

Marine animals follow the *tidal* rhythm, which is established by the regular pull of the moon's gravity on the Earth's waters. Sea animals use the *lunar* rhythm, which brings very high and very low tides once each month, to regulate their breeding cycle.

Circannual Rhythm

Circannual rhythms are patterned according to the year. The seasonal activities of animals–reproduction, hibernation, migration–are governed by circannual rhythms. So are the sprouting, flowering and seed production patterns of plants.

Plants and animals have followed the Earth's natural rhythms since life began, passing on the unique rhythmic patterns of their ancestors from generation to generation. These are not learned rhythms; they are innate. They are present when a seed drops to the ground and when a baby is born. Even tiny chicks, hatched in an incubator and without the guidance of a mother hen, know to eat when the light is on and rest when the light is off.

Fast Fact: The tiny "pineal gland," found in the brains of all vertebrate animals is known as the "master clock" of all light-dependent activities. It sets the sleep/wake cycle in mammals and seasonal rhythms in mammals and birds.

Fast Fact: Chronobiology is a new branch of biology that deals with biological rhythms and the important role time plays in ordering life processes.

Don't Mess with Mother Nature's Watch

Even when you are not aware of changes in the light outside, your body follows a "circadian" or day/night rhythm. This body clock is set off about 25 hours, which is very close to the 24 hours, 50-minute rhythm of the tides. In scientific experiments, volunteers were kept in an underground room on a 28-hour schedule. They did not adjust well to the new sleep/wake pattern and felt as if they were suffering from "jet lag." Jet lag, or *circadiandysrhythmia* or *desyncrhonization* is a feeling of unusual tiredness and confusion experienced when you fly a great distance east or west across the Earth's time zones. The sudden changes in day length puts your body clocks out of order and your natural rhythms out of sync with the world around you. It's hard on the body and brain and can take several days before your internal clocks are reset.

Experiments have shown the importance of the day/night cycle on living things.

When wintering silkworm cocoons were exposed to 16 hours of continuous light (the same amount of light they would receive in the middle of summer), the pupae inside stopped developing. Other pupae, who were only exposed to eight hours of daylight (the correct amount of light for a winter day), continued to develop.

When tiny birds called juncos were captured on their migration south from Canada and kept in an aviary that was cooled to 32°F (0°C) (the average temperature during a Canadian winter) but exposed to more than the usual amount of daylight, they were singing and getting ready to start their families mid-December! When they were released into the winter outside, they headed north instead of south, back toward Canada and their spring breeding grounds.

Birds become confused during an eclipse of the sun. The sudden darkness disrupts their timekeeping, and they become restless, thinking that they should roost but knowing that it is the wrong time.

Fast Fact: It is not only humans that associate the full moon with danger! Some small animals live their lives differently when the moon is full. Because the night is so bright, they alter their normal rhythm of activity and stay quiet, hidden and safe from their sharp-sighted nocturnal enemies for most of the night.

Fast and Fishy Fact: Even shallow-water fish seem to know the difference between day and night, acting differently according to whether it is light or dark. (Deep-sea fish have no knowledge of day and night since no light reaches them in their dark world five or six miles (8 to 9.5 km) below the surface of the water. Scientists suspect that these fish are unaffected by the day/night cycle.)

Body Clocks

You probably don't realize it, but you have a vast collection of clocks inside you! These body clocks are not the regular sort of clocks. You can't see them; you can't feel them; and you can't change them. They are nature's clocks, and they are a gift. You don't even have to know how to read them. Your brain does that for you.

Together, these clocks are what keep you alive. Our internal clocks tell us when to sleep, when to wake up, when to eat, when to go the bathroom, when to work and when to have fun. In many ways, we are, and always have been, slaves to time.

Among the clocks you keep tucked away in your brain and bodies are the sleep/wake clock, the body temperature clock, the breathing clock, the kidney clock, the pulse clock, the winter/summer clock, the cell clock, the eating clock, the immune system clock, the blood pressure clock, the hormone clock, the blood sugar clock and the reproduction clock.

These clocks are set at different times, but keep the same time day after day. To a certain extent, we can adjust some of our schedules–waking, working, eating and sleeping, for example–but we can never stray too far. No matter how determined we are to stay awake, at some point awake time runs out and we fall asleep.

The Sleep/Wake Clock

We all need sleep. Why this is so is not entirely clear, but one thing is certain: we cannot live without it. When our sleep/wake clock says it is time to sleep, we sleep. When it says it is time to wake up, we wake up.

Even though we are in a state of near-unconsciousness while we sleep, important things are happening: our bodies are being restored; our tissues are repairing themselves; we are growing; and our cells–in the top layer of the skin, in the lining of the mouth, in the eyes and in the hair follicles–are being renewed. And of course, we are dreaming!

The amount of time we spend dreaming depends on how old we are. The deepest dream state of sleep (REM or rapid eye movement) accounts for 18% of the sleep of children aged 21 months to 19 years, 13% of the sleep of people aged 20 to 29 and only 3% of people aged 50 to 69 years of age.

Fast Fact: Even our sleep time is on a schedule. Each night we pass through four or five cycles of dream and nondream sleep, with each complete cycle accounting for 80 to 100 minutes of our full night's slumber.

Good Times/Bad Times

There seems to be small differences in our "alertness" throughout the day. We generally feel "smartest" in the morning, and this is when we are best able to remember things, figure out math problems and–believe it or not–play video games! Our brain skills tend to peak in the middle of the morning, drop in the afternoon and bottom out in the middle of the night (which is another reason why we sleep at night and move about during the day).

Daylighters and Nightlighters

Like many other living things, human beings are diurnal. This means that we are active in the day, and we sleep at night. This could be because we are not too capable in the dark. Sight is important in almost everything we do, which means we are most efficient during the daytime.

Some living things are nocturnal—that means that they are specially designed for nighttime activity. Their clocks are set opposite to ours for a reason. These creatures have adapted in some way—owls with night vision, bats with radar, for example, to help them hunt and survive without light.

The following pictures are both the same. It's your job to make them different! Make one picture show a nighttime scene and the other a daytime scene.

Night   Day

What Kind of Bird Are You?

Even though our bodies all contain the same clocks, each of these clocks is programmed to a slightly different schedule. Just as all people do not work to the same day length (some day/night clocks run at just over 24 hours, others at almost 25) our patterns of alertness also vary. Some people, the ones who get off to a roaring start but wear out later on, like to go to bed early and get up early. Others, those who function best in the afternoon, like to stay up late and sleep in. These two types of people are often referred to as "larks" and "owls." As expected, larks find that their temperature and pulse peak in the morning, while owls peak in the afternoon. The majority of people, however, are neither larks nor owls but some kind of bird (so far unnamed) that falls right in the middle.

Are you a lark, an owl or a no-namer? To find out, keep a diary over a seven-day period. At regular intervals each day–8:00 a.m., 12:00 noon, 4:00 p.m.,?–record your mood, energy level, physical strength, attentiveness, ability to learn and temperature. In each area, rate yourself on a scale of one to five as follows:

Mood:	1. blue	2. blah	3. normal	4. happy	5. elated
Energy:	1. lethargic	2. sluggish	3. normal	4. energetic	5. bursting
Strength:	1. no strength	2. weak	3. normal	4. fairly strong	5. Herculean
Attentiveness:	1. unfocused	2. wandering	3. normal	4. quite focused	5. right on
Receptiveness:	1. no go	2. a little	3. normal	4. picking up	5. getting everything
Temperature:	1. below normal		3. normal		5. above normal

Tally your scores at the end of the week. If your average score is in the 3 range in all areas, or if you score high in the middle and low on either side, you are probably a no-namer. If you tend to score high in the morning and lower in the afternoon, you are probably a lark. If, on average, you score low in the morning and high in the afternoon, chances are you are an owl.

The Cellular Clock

Within your body, different cellular clocks tick at different rates. Liver cells divide every year or two, while the cells that line your stomach divide twice each day. Cells from a human baby divide 50 times before they die, while those from a 40-year-old divide just 40 times, and cells from an 80-year-old divide only 30 times. This cellular slowdown suggests that living creatures are programmed to live for a fixed amount of time.

Body Facts

The body cells with the shortest life span are those in your stomach. They are shed after only three days. The longest-living cells are those in the brain. They last you for life–up to 90 years!

Your body sheds about 50 million dead skin cells every day!

Your fingernails grow at a rate of about 0.02 inches a week–four times faster than your toenails. If you did not cut them for a whole year, they would grow to be a little more than 1" (2.5 cm) long.

Your hair grows about half an inch each month, but it doesn't grow forever. It usually stops after two or three feet.

The Temperature Clock

Unless you are ill and have a fever, your brain keeps your body at a fairly constant temperature of 98.6°F (37°C). But there is a daily rhythm of body temperature that moves your internal thermostat up and down by as much as 1°F (0.5°C). Your temperature is lowest first thing in the morning, just before you wake up, and is highest at around 9:00 at night.

Because cold-blooded animals set their body temperature according to the temperature of their surroundings, they are at the mercy of the day/night rhythm and the regular cycling of the seasons. Warm-blooded animals, on the other hand, have a little more freedom to do as they please. This is because of a survival strategy called homeostasis, which comes from the Greek words *homeos* meaning "similar" and *stasis* meaning "state" or "condition." Homeostasis allows us to keep conditions inside our bodies relatively stable, and it is nature's way of making time a little less important. The internal thermostats of warm-blooded creatures fluctuate to keep their bodies at relatively the same temperature inside regardless of the chill or heat outside. (And, of course, when the changes in external temperature are too extreme, you can adjust your clothing or retreat to your home, where fireplaces and furnaces, fans and air conditioners do most of the work for you!)

Fast Fact: People seem to perform their best at the time of day when their body temperature is highest. Why? Possibly because the brain metabolizes fastest at peak body temperature.

The Heartbeat Rate Clock

Your heartbeat also follows a rhythmic pattern, and this particular clock is set according to how old you are, what you are doing and what you are thinking. At rest, a baby's heart beats about 120 times each minute, while a child's heart beats only 90-100 times per minute and an adult's heart beats a mere 70-80 times each minute. When you put all these together over the course of a lifetime, the average heart beats five billion times! At night, when your body is at rest and doesn't need as much oxygen in its bloodstream, your heart beats more slowly. During the day, when you are alert and active, your heart beats more quickly. The more active or anxious you are, the faster your heart beats.

Take Your Pulse

You can see how fast your heart is beating by counting the thumps you feel in your chest or the little bumps you feel in your wrist. Here's how to do it.

At Rest
- Find a quiet place to sit.
- While you are resting, find your pulse, either at your chest or wrist.
- Look at the clock.
- When the second hand reaches 12, start counting.
- Stop counting when the second hand reaches 3.
- The number of thumps or bumps you felt was how many times your heart beat in 15 seconds.

At Play
- Run around or jump up and down for a few minutes.
- Sit on the floor and take your pulse again.
- How has your heartbeat rate changed?
- Hold your breath for a few seconds.
- Does your heartbeat rate change? Why?

Quick Query: What would your heartbeat rate be like after a nightmare?

Animal Heartbeat Rates

In warm-blooded animals, heartbeat rates are generally linked to body size: big animals have slow heartbeat rates; small animals have faster ones. Heartbeat rate is also linked to life-style. Creatures that use a lot of energy running or flying need to have faster heartbeat rates than animals who move slowly. In cold-blooded animals, heartbeat rates vary according to temperature. The cooler the temperature, the slower the heart rate.

Take a look at these heartbeat rates: (b/m = beats per minute)

Shrew: 800 b/m	Seal: 100 b/m
Bat: 660 b/m	Ostrich: 65 b/m
Starling: 400 b/m	Crocodile: 30-70 b/m
Cat: 200 b/m	Elephant: 30 b/m

Fast Fact: Galileo Galilei, made his discovery of the pendulum by using his pulse to time a swinging chandelier.

The Breathing Clock

Breathing is another natural and essential rhythm. All living things breathe continuously from the day their lives begin until the moment they die. You take a complete breathe-in and out-about 12 to 15 times every minute. And this pattern is regulated by another body clock which is automatically set by your brain. For the most part, we breathe on "autopilot." This clock, too, is timed to your activity level. When you are asleep or at rest, you take slow deep breaths; when you awake and active, you take rapid, shallow breaths.

Fast Fact: If you live to be 75 years old, you will have taken about 570 million breaths!

It is possible to momentarily override your natural breathing clock. For example, you can probably hold your breath for about one minute-especially if you take some fast, deep breaths first to build up your internal oxygen supply. But it isn't long before your brain regains control and resets the rhythm.

 ## How Many Breaths?

Find out how many breaths you take each minute.

Materials

timing device with a second hand
you and a friend

What to Do

1 Have one partner hold the timing device and the other prepare to breath with as natural a rhythm as possible.

2 The timekeeper tells the "breather" when to start counting breaths and when to stop.

3 Record the number of breaths taken in one minute.

4 Switch roles and record the number of breaths of the other partner.

5 Compile breathing statistics from as many participants as you can.

6 Compare the number of breathes in a mathematical exercise. Who took more breaths? Who took less? How many more breaths did ___ take each minute than ___? What is the average number of breaths taken each minute? If you take ___ breaths per minute, how many times would you breathe in an hour? In a day? How many seconds are there between breaths?

Fast Fact: The world record for breath holding is six minutes!

Fast Fact: The world's best breath holders are diving mammals. The breathing clocks in whales slow the giant animals' heartbeat rates while they are underwater and shut off blood supply from all but the most essential organs to conserve oxygen. In this way, whales can hold their breath for more than 30 minutes while deep-sea diving!

As the Stomach Churns . . .

Your stomach and the tube that leads out of it (the intestine) contain muscles that contract and relax according to regular rhythmic pattern. The motion of these muscles, which starts at the back of your throat the moment you swallow, pushes whatever you have eaten through your digestive system. A complete trip-in one end and out the other-takes between 12 and 24 hours.

The Speech Clock

Even our speech and everyday conversation follow a regular rhythm. In fact, the time-keeping mechanisms of speech are so well refined that language would be impossible without them. The speech sounds that we use have a definite order in time. We cannot call out for help, for example, if we mix up this order, or the timing of each sound. If you were to yell out, "PLEH!" not too many people would come running. Similarly, if you hollered HE! and then waited a couple minutes before adding the LP! your chances of a rescue would be slim.

This unconscious and precise timekeeping is essential, too, to the success of a conversation, where speakers and listeners interact with one another in a kind of concert. One word spoken too early or too late can ruin the whole thing. To be an effective partner in conversation, you have to be able to "seize the moment" in the flow of talk. Just as timing is everything for the animal who will either eat or be eaten, so it is with conversation. If the timing is right, you will be able to talk meaningfully to your friends. If it is wrong, you will just walk away frustrated.

Analyze a conversation. Take notice of the pauses and the time people leave between words. This is done so people can breathe and talk but also for effect-to help make their conversation more meaningful.

The Sick/Well Clock

The more we learn about our internal clocks, the better we will be able to diagnose illness and treat people who are sick. Studies have already shown that there is a good time and a bad time to administer medications and special treatments, but "normal" routines-when people take medicine, when meals are served-do not take biological rhythms into account. One study found that blood pressure medications are much more effective when taken at 6:00 p.m. and midnight than at any other time. And in laboratory mice diseased with cancer, the timing of chemotherapy and radiation treatments meant the difference between life and death. Ninety-six out of 100 mice died when treated with chemotherapy at 2:00 in the morning. Only four died when treated with the same dose of chemotherapy at 8:00 in the morning or 5:00 in the afternoon. For human cancer patients treated with chemicals or radiation, receiving treatment at the right time may mean the difference between getting better or worse. Determining the settings of the cancer cell clock-when cells grow and divide, peak and drop-may give us a clue as to when they are most vulnerable to drugs or radiation.

How's Your Timing?

K-4 • Language/Dramatic Arts • Communication/Public Speaking/Creative Thinking • 10 min

Some people have what is called good "timing." They are good at keeping time in music or dance, they show up just in time for good opportunities or they know just how to deliver a joke to make people laugh. Try your "timing" with the following jokes and lines:

What time is it when an elephant sits on your fence?

Time to get a new fence.

What did the hands say to the clock at bedtime?

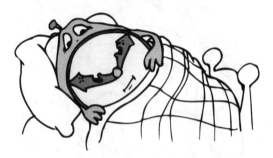

Stop "tocking."

Why did the boy throw the clock out the window?

To see time fly!

What kind of dog keeps the best time?

A watchdog, of course!

Try using these lines at just the right time:

Yours until the ocean waves.
See you when the lipsticks.
I'm yours until the milk shakes.
Yours until the earthquakes.
Yours until Niagara Falls.
Good-bye until butterflies.
Yours until the kitchen sinks.
I'll be back when the artichokes.

114

The Body Clock of _____ Works Like This

What schedule are your body clocks set to? When do you sleep? When are you awake? When do you feel like being active or quiet? When do you feel like eating?

Fill in the following spaces with the activities your body clocks dictate throughout the day.

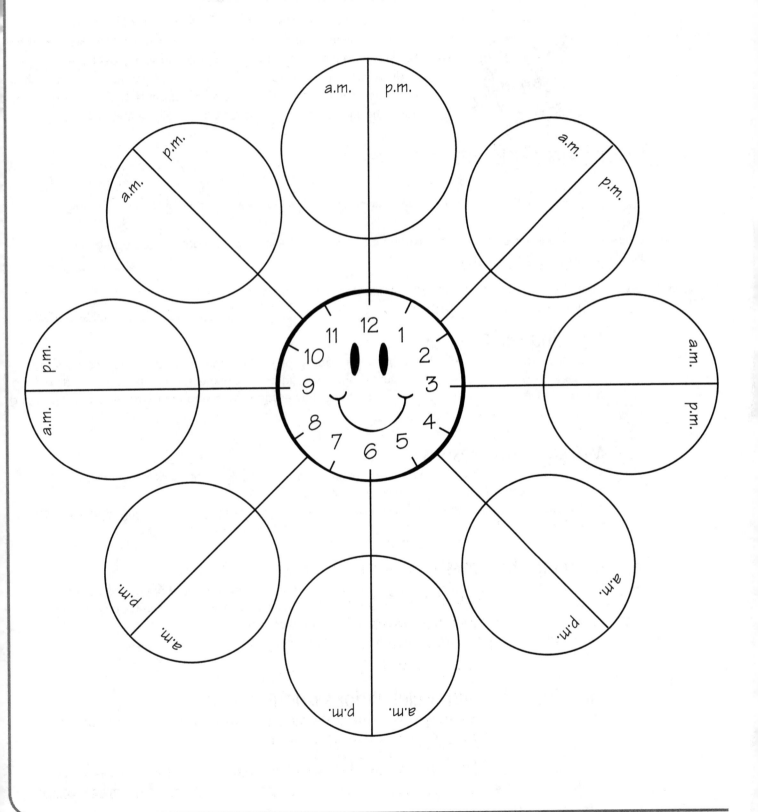

Chapter 9
Mother Nature's Timekeepers

Nature's Clocks

We depend on the rotation of the Earth for our 24-hour clock, the moon's rotation around the Earth for our months and the orbiting of the Earth around the sun for our solar year. Did you know there are an endless number of natural clocks in the world around us? Nature has provided us with many timekeepers from animals, birds and insects to tides and plants. Look around and try to tell the natural time.

Marking Time with Nature

K-4 • Science/Language/Math • Observation/Fact Recall/Listening/Counting • 30 min

Early people relied on the sun, the moon, the stars and other natural phenomena to mark the passage of the time. Discuss natural clues which signal the passage of time.

Quick Query: *What clues do you notice that indicate the passage of time?*

Marking the Day
What signs do you notice that help mark the passing of time throughout a day? Lead children to recognize the following signs: sunrise, sunset, shadows, light and darkness, a rooster crowing, dew evaporating off the grass, plants opening and closing, flower faces following the movements of the sun and so on.

Marking the Seasons
What do you notice that helps you measure time with the passing of seasons? How many springs, summers, winters or falls have you seen in your lifetime?
Help children to recognize the following as signs of seasons changing and passing: buds, blossoms, fruits and harvest; weather changes; falling leaves; migration patterns and so on.

Marking the Years
How do you mark the years using signs in nature? Have you ever thought of yourself as a natural, growing, living thing? Can you mark the passing years using yourself as a time-keeping device?
Draw children's attention to their own growth and physical changes.
Could you always walk and talk and swim? Do you know any old trees, old people or perennial plants that increase in size every year?

Marking Time with Celebrations and Festivals
Most customs and traditions arose from early celebrations that were connected to natural cycles of the Earth, the heavens and the harvests required for existence.
What do you see in nature that could be celebrated?
Take note of the spring rains, the first greenery, the summer sunshine, the fall harvest, the arrival of wildlife and the first snow of winter. Annual celebrations provided people with a way to measure the passage of a year.

The Tides of Time

Tides are the rhythmic rise and fall of water in the Earth's lakes, oceans and seas that occur twice every day. The culprit? The moon and gravity.

As the moon makes its daily journey around the Earth, it pulls on the land and water that is directly below it. The solid land doesn't move, but the water does, bulging up in the middle. At the same time, the water on the opposite side of the Earth is bulging because the Earth itself is being pulled toward the moon and the water hangs behind. The two bulges follow the moon and move around the spinning Earth.

At low tide, the water is pulled away from the beach, leaving lots of room for sand castles. At high tide, most of the beach is covered with water. Every 24 hours (and 50 minutes), there are two high tides and two low tides on every beach around the world. Because the interval between high tide and low tide is always the same, the time of day when both of these occur changes every day, following its own rhythmic cycle.

Very high tides, called spring tides, occur when the sun, moon and the Earth line up. Low neap tides occur when the moon and sun are at right angles to one another.

Fast Fact: When it is high tide on one shore, it is low tide on the opposite shore. When the North American shore of the Pacific Ocean is experiencing high tide, the Asian shore is experiencing low tide.

 ## Tide Craft

K-1 • Science/Art • Concept Reinforcement/Creativity/Fine Motor Skills • 30 min

Materials

sturdy construction paper or tagboard
white craft glue
paintbrushes or glue sticks
sand, shells, pebbles, twigs and other seashore items
large mouth plastic container for sand
markers

What to Do

1 Have children plan a seashore scene and then color the "water" and "sky."

2 Paste can be spread on the "shore" area.

3 Pour sand over the glue and gently shake the paper back and forth to spread the sand. Pour remaining sand back into the original container.

4 Other natural objects can be glued to the sandy surface.

5 Allow the creation to thoroughly dry before moving.

Discuss ways in which humans and sea and land creatures are affected by the tides.

117

Seasonal Time

No matter where you are in the world, the seasons officially begin on the same day. Spring begins on March 21, summer on June 21, fall on September 21 and winter on December 21. Although the dates are the same everywhere, the seasons can be quite different from one place to another.

Mark the Arrival of the Seasons

The official calendar dictates the days, months, years and the seasons–but the natural world of humans, animals, birds, insects, trees, plants and weather lets you know when a season has naturally arrived. You don't need a calendar to tell you that the seasons have changed–you can look to nature. Birds foretell the coming of spring, squirrels prepare for winter without a calendar to tell them the dates and the falling snow usually lets you know that winter has really arrived. Take notice of and mark the natural changes that each season brings. On what date does the season naturally arrive?

 ## Chart the Natural Signs of the Seasons

K-4 • Art/Science • Reinforcement of Seasonal Concepts/Drawing/Group Interactions • 30 min

Materials

mural paper
markers, crayons,
   pencils, erasers
magazines
scissors
paste

What to Do

1 Divide the mural paper into four equal sections. Mark the sections: spring (March 21–June 20), summer (June 21–September 20), fall (September 21–December 20), winter (December 21–March 20)

2 Divide the children into four equal groups and allow each group to choose a season.

3 Have each group design and decorate their mural area to represent their assigned season.

Try This

- Begin the exercise by doing some imaging exercises. Have children picture themselves in spring for instance. What are they wearing? What are they doing? Can they smell anything? What do they see? What do they hear?

- **Record and Mark the Natural Time:** Make tallies, diaries, charts, posters, murals, time lines, circular dials or other creative projects to take a look at time passing in the natural world.

Throughout history people have relied on superstitions, wisdom and sayings—some arose from wisdom and some from folly—to know when to plant and harvest their crops. Do you think the following are based on wisdom or foolishness?

- When the dove sings, winter is over and planting time is on its way.
- Always plant beans in the morning.
- Plants that will grow above ground should be planted at sunrise to rise up with the sun.
- Plants that grow into the Earth should be planted at sundown to go down with the sun.
- All is safe for planting after May 24th.
- Plant sweet potatoes on April 8.
- Flowers should be planted in May.

Plant Time

Each kind of plant grows according to its own clock or calendar, and follows the same cycle, year after year—winter dormancy, spring awakening, summer growth and autumn fading.

Plants set their seasonal "flowering" schedule by the amount of light they receive during a day. The "photo period," or light requirements for blossoming differ from plant to plant. Some plants need a short day to flower, others a long one.

To produce seeds and future generations, plants must grow and flower when conditions are just right. This means that plants must be able to predict what will happen next. They must be able to "tell" time. How do seeds know when to start growing?

During the winter, seeds lie dormant underground, their tiny, newborn leaves protected from the frost and the bitter cold by a tough jacket or seed case. Nature has set their timers in such a way that there is a period of delay built into their growth cycle. When this period of delay is over and the days get longer, the seed case splits open and the plant begins to push skywards. After that, it's up to the leaves.

The plant relies on its "aerials" or green leaves to tell it when to start the flowering process. The leaves can sense changes in the amount of daylight and pass on this information (in the form of chemical signals) to the rest of the plant. When day length is just right, the plant produces buds.

A Time to Grow

The length of the growing season – the time when plants grow, flower and die – is different around the world. In temperature climates, the growing season is relatively long, but the amount of daylight available changes as spring gives way to summer and summer to fall. Different flowers bloom in different seasons according to their particular photo periods. In the tropics, day and night are of equal length all year long, and flowers may bloom more than once a year. In Arctic lands, such as Siberia and parts of Alaska, the sun never sets during the two months of summer. In these places, plant clocks are set to 24 hours to take advantage of the nonstop sunshine.

On your mark, get set, FLOWER!

Plants depend on day length and water to flower. In arid deserts where rain is a rare occurrence, the landscape appears brown and lifeless for much of the year. When the rain does come, the desert comes alive overnight. As if a magician has waved a magic wand, it seems to explode with flower, and for a few short weeks the desert is a garden of lush and rare beauty.

Why the hurry? Desert plants are timed to rush. Their clocks are set to produce flowers and seeds quickly before the rain stops and the water dries up. The insects that feed on these flowers also have rapid-action clocks. The brief visits of these newly awakened flies, beetles, bees, moths and butterflies fertilize the flowers and ensure that there will be a new crop of seeds lying in wait for the next rainstorm.

Fast Fact: Sunflowers can grow from tiny seeds to plants taller than a man in under six months. Climbing plants like the Russian vine can grow even faster, reaching 20 feet (6 m) in six months.

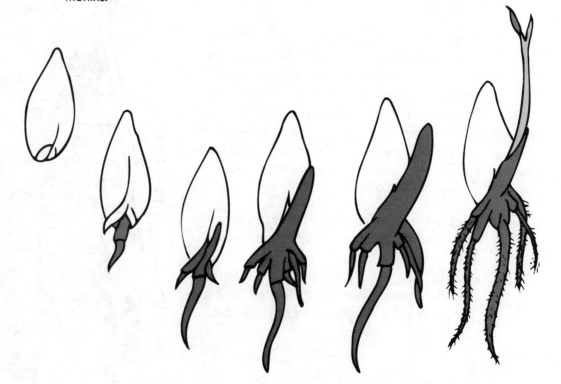

Spring in Winter

You can change a flower's clock, so to speak. By knowing the photo period of a plant and artificially extending or shortening the length of day, flower growers can make greenhouse plants flower when they want them to. Timing things just right, they can produce lilies for Easter, chrysanthemums for Thanksgiving and poinsettias for Christmas even though it is winter outside.

To make a plant flower in time for Easter, greenhouse gardeners shine light on lily seedlings all night long during the months of December and January. Because the day is now 24 hours long, the seedlings grow fast, producing tall stems and lots of leaves. But they will not flower in continuous light. So, in March, when the days are still short, the gardeners turn the night lights off, and the plants quickly produce flowers – just in time for Easter.

You can do the same thing in your classroom. Try the following activity to witness firsthand the effects of day length on plants.

Materials

4 bean seeds
2 glass jars, labeled 1 and 2
paper towels
water
light that can be left on 24 hours a day
dark cupboard or box

What to Do

1 Line the glass jars with paper towels. Label one *natural* and the other *light*.

2 Place two bean seeds in each of the two glass jars, making sure that the bean is located about halfway up the jar between the glass and the paper towel.

3 Put a little water in the glass jar. (The paper towel will draw the water up and keep the bean seeds moist.) Make sure there is always a little water in the jar.

4 During the day, put both jars under the light.

5 Before you leave the classroom, put the "natural" jar in the cupboard or box. Leave the "light" bean under the light.

6 After about a week your beans should start to grow. Which beans grow the quickest? What differences do you see between the two jars? What affect does the 24-hour day have on the seeds? What about the regular day? What do you think would happen if the beans in the one jar never got a break from the light?

Try This

- Carry out the experiment long enough to find out what would happen if one bean plant is never exposed to darkness. Transplant the beans into two pots of soil and keep up your pattern. What happens to the beans in each pot?

121

Flower Clocks

Depending on what conditions they require for growth and what insects they require for pollination, plants open and close at different hours of the day. Carolus Linnaeus, the 18th century botanist, was the first to recognize this and made up a flower clock that would allow people to tell time according to when certain flowers opened and closed. Linnaeus clock was so accurate that for years afterward Europeans planted flower beds in the shape of a clock, each bed blossoming at its own hour. On a sunny day, they could tell the time just by looking in their garden.

This was Linnaeus's clock:

   6:00 a.m. Spotted Cat's-Ear (opens)
   7:00 a.m. African Marigold (opens)
   8:00 a.m. Mouse-Ear Hawkweed (opens)
   9:00 a.m. Prickly Sow Thistle (closes)
   10:00 a.m. Common Nipplewort (closes)
   11:00 a.m. Star of Bethlehem (opens)
   12:00 noon Passionflower (opens)
   1:00 p.m. Childing Pink (closes)
   2:00 p.m. Scarlet Pimpernel (closes)
   3:00 p.m. Hawkbit (closes)
   4:00 p.m. Small Bindweed (closes)
   5:00 p.m. White Water Lily (closes)
   6:00 p.m. Evening Primrose (opens)

Another clock, invented some time later, uses a different blend of flowers and starts later in the day, telling time by the opening of petals only.

   8:00 a.m. Pimpernel
   9:00 a.m. Field Marigold
   10:00 a.m. Alpine Dandelion
   11:00 a.m. Star of Bethlehem
   12:00 noon Passionflower
   1:00 p.m. Carnation
   2:00 p.m. Squill
   3:00 p.m. Pyrethrum
   4:00 p.m. Purple Hawkweed
   5:00 p.m. Catchfly
   6:00 p.m. Evening Primrose
   7:00 p.m. White Lychnis

Make a Living Flower Clock

Materials

plot of earth
fertilizer
seed catalog and seeds
garden spade, trowel, rake, hoe
watering device
sunshine and rain
chicken wire or other fencing material if necessary

Get Started

1 Contact your local horticultural society or nursery to find out what time different flowers open and close in your area.

2 Once you know which flowers open at what time, get your hands on a seed catalog. They always have beautiful pictures of flowers in full bloom.

3 Draw a circle and divide it into 12 equal parts, one part for each hour on your flower clock.

4 Design your flower clock. Decide what will be the first hour on your clock (you do this by determining which flower opens first).

5 Then, either copy and color or cut and paste the flowers onto the clock, according to their hourly schedule. Labels may be necessary.

6 Now that you know what the different flowers in your clock look like, start looking for a garden plot.

What to Do

1 Prepare the flower bed with your garden tools and fertilizer as early as you can break ground. Design the plot with each seed variety in its own section.

2 When your seeds arrive, read the packages carefully and consult other sources–garden professionals or resource books for details regarding the planting of the particular seeds.

3 Assign various groups of children to plant the seeds as directed in each of the 12 clock areas.

4 Water and wait!

5 Check on your garden at different times of the day. Can you use your flowers to tell the time of day?

Try This

• Provide seed catalogs, magazines, scissors, paper and paste so children can create flower clocks on paper.

• Take a close look at the next flower bed you see. Can you tell the time just by looking at the flowers?

123

Creature Clocks: Travelers, Sleepers and Tough-Guys

In temperate climates such as Europe and North America, there are drastic changes in temperature between the summer and winter seasons. Animals need more energy to keep their bodies warm in the cold, and this means more food. But food supplies are scarce in winter. Plants stop growing. Snow covers feeding grounds. Water turns to ice. Every creature adapts to these changes in its own way.

Think of the birds and animals in the area where you live. Make a list of travelers–those that migrate to warmer winter homes; the sleepers–those that hibernate for all or part of the winter; and the tough-guys–those that make the most of what they have and just change their clothes to suit the weather!

People are tough-guys. Make a list of your winter preparations. What happens when the weather gets warmer in the spring?

Migration

For some animals, the solution to the winter problem is to migrate, or travel, to warmer places for the coldest months and then return when the weather is warmer and the food more plentiful. These air, sea or land journeys are often very long and require great strength and endurance.

Many species of birds fly south in the winter, often travelling in flocks to minimize the threat of danger. The birds are able to set off together at both ends of their journey by telling time through changes in temperature and daylight hours. Monarch butterflies also have a strong built-in "clock." They spend the summer months in Canada but fly south when the weather turns cool in the fall, travelling more than 1,800 miles (3,000 km) to gather by the billions in wooded areas in Florida, Texas, California and Mexico.

Animals migrate for other reasons, too: to look for fresh supplies of food, to find a safe place to lay their eggs or to give birth to their young. Caribou spend the winter in the forests of southern Canada but travel some 800 miles (1,287 km) to the far north to feed on the lush greenery that grows in the short Arctic summer. Other animals, like salmon and green sea turtles, travel great distances to lay their eggs in the exact places their ancestors visited hundreds of years before. Salmon, swim *up* waterfalls to get back to the shallow streams in which they were hatched, and green sea turtles travel 1,400 miles (2,253 km) from their feeding grounds off the coast of Brazil to mate and lay their eggs on the very same isolated Ascension Island beaches where they emerged.

Fast Fact: The Arctic tern makes the longest migration trek of any creature in the world, traveling from the top of the world (Arctic) to the bottom (Antarctica) and back again every year. It's a 9,000-mile (15,000 km) journey as the crow flies.

124

Hibernation

Some animals escape the rigors of winter by "hibernating." Skunks and certain squirrels, for example, settle into a deep sleep safe in holes, burrows or caves. While hibernating, the body clocks of these animals slow down dramatically to conserve energy and precious supplies of body fat. Their temperature drops, their heart beats more slowly and they breathe shallowly and less frequently.

Bears, badgers and raccoons also spend the worst of the winter months asleep, but they are not "true" hibernators since their temperature and breathing rates do not really change, and they wake up during mild spells to forage for food.

Fast Fact: The longest hibernator in the world is the Arctic ground squirrel of northern Canada and Alaska. It lies dormant for nine months of the year!

Adaptation

Other animals do not hide away but grow thicker winter coats to keep their bodies warm in the cold. These animals change their wardrobes according to the day/night clock. Their winter coat grows when daylight hours grow short in the autumn. When the day lengthens in the spring, the winter coat is "shed."

Animals change their coats in two different ways. Some simply change the color of their fur. Dark summer fur blends with grasses, while white winter fur provides camouflage in the snow. Others grow thicker, warmer coats to protect themselves against the bitter cold.

Tiny Timekeepers

Insects set their clocks according to what is going on in the world around them. Like most other clock watchers, the day/night cycle and length of daylight are the most critical settings. Butterflies, for example, are active during the day, while moths are active at night. And all insects time their growth, reproduction and death, according to the calendar.

Pest Predictions

Farmers and fruit growers can control pest population by understanding and interfering with the internal clock settings of their enemies. For example, cabbage farmers know that cabbage worms, the destructive larva of the white cabbage butterfly, make their cocoons in the short days of early fall. By shining lights on the cocoons at night to artificially extend the hours of daylight, the farmers trick the worms into thinking that spring has come. Because they use length of daylight as a growth trigger, they stop developing and prepare to emerge from their cocoons. When spring finally does come, the worms do not emerge as butterflies, and the farmer is rid of the pest for at least one year.

Fast Fact: Honeybees are awesome clock watchers and "tell" time by the position of the sun. Their timekeeping is so accurate they can return day after day to the same flower at the exact moment it opens its petals!

125

Nature's Diaries

Not only does nature supply living things with built-in clocks, nature creates diaries, or records of times past, that can be studied and learned from in the future.

Tree Time

The age and life story of a tree are indelibly inscribed in its trunk and can be discovered by examining the number of double-banded light/dark growth rings in a smooth-cut stump. A tree's trunk is made of microscopic wood tubes, called *xylem vessels,* that take water and minerals from the roots and distribute them to the rest of the tree. A new layer of these tubes is always growing just beneath the bark. In the spring and early summer, growth is fast and the new tubes are wide. This is the light half of each ring pair. In the late summer and early fall, when growth slows down, the tubes that grow are thinner, and these form the dark half of the pair. The number of light/dark ring pairs tells us how old the tree is. Each pair represents one year's worth of new growth. The relative thickness of the rings tells us if the tree had a good year or a bad year. Thick ring pairs indicate that the weather was good for growing trees. Many xylem vessels were required to take up plentiful water and food supplies. Thin ring pairs indicate dry or cold weather and the need for fewer, smaller wood tubes.

Fast Fact: These days, experts don't have to cut down a tree to see how old it is. They can bore from the outside through to the center and take out a cylinder of wood that shows the complete sequence of rings.

Fast Fact: The oldest trees in the world are the slow-growing bristlecone pines found in California. Some of these trees are more than 4,000 years old, making them the oldest living things on Earth. Some of the dead bristlecone pines that are still standing began their lives more than 8,000 years ago.

Read the Diary

Take a walk in the woods or to a community spot where a tree has been cut down. It is easiest to read if the stump has a nice flat top. Pour a little water on the stump to make the rings more visible. Start at the center of the stump and count the pairs of rings.

How old was the tree when it was cut down? How many good growing years did it have? How many poor growing years?

Reading a Tree Diary

Look at the tree trunk in the picture at the right. Can you tell:
- how old the tree was?
- how many good years it had?
- how many bad years it had?

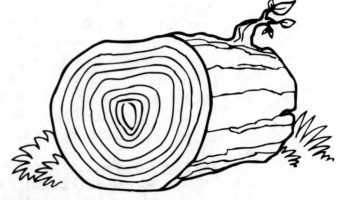

Tortoise Time

Because the number of hard, bony plates that make up a tortoise's shell does not increase as the turtle gets older, the plates themselves grow-like growth rings-at different speeds during the year depending on weather conditions and food supplies. By counting the yearly growth rings on one plate, you can tell how old the tortoise is. Unfortunately, this method only works well for tortoises up to the age of 20 or so. When a tortoise gets very old-and they can live for more than 100 years!-their shells become worn and scratched, making it almost impossible to see the rings clearly enough to make an accurate age count.

Antler Accuracy

Antlers are living bone that grows, sheds and regrows according to a yearly rhythm. This annual cycle is controlled by a natural rhythm built in to the animal's body which is triggered by the length of daylight. Male deer, moose and elk need antlers to fight for females so each spring or summer they grow a new set of antlers. The velvety fur on the antlers is shed in the fall, but the bare-boned antlers remain alive until after breeding season is over, and the antlers die and drop off during the winter. Each time the cycle starts over, the antlers grow longer and are more branched than before.

Carbon Dating

Scientists can estimate the age of ancient remains by measuring the carbon-14 present.

Rock Clock/Diaries

Using a process called relative dating, paleontologists can figure out how old a fossil is and when a creature lived from the age of the rocks around it. It takes from thousands to millions of years for sediments laid down in the sea to form thick layers of rock. These layers help us to determine the prehistoric period of a particular item.

Fossil Time

Prehistoric Stages	Years Ago		
Quaternary period	2,000,000	-	today
Tertiary period	65,000,000	-	2,000,000
Jurassic and Cretaceous periods	195,000,000	-	65,000,000
Triassic Period	230,000,000	-	195,000,000
Carboniferous and Permian periods	345,000,000	-	230,000,000
Devonian period	395,000,000	-	345,000,000
Ordovician and Silurian periods	500,000,000	-	395,000,000
Cambrian period	570,000,000	-	500,000,000
Precambrian period	4,000,000,000	-	570,000,000

Fast Fact: The famous Sunday Stone is a permanent unwritten record of human history. This rock diary was formed in an English coal mine during the 1800s. A white mineral, called barium sulphate, settled in a water trough, and during working shifts was blackened by coal dust. From the pattern of bands that was formed, we can tell that the miners worked six days a week and rested on the seventh.

127

Chapter 10
Time in a Bottle

Time Talk

It is amazing how often we use the word *time* in our everyday language. Think about it the next time you listen to a story, see a newspaper or watch a show on television. Is it possible to do anything without involving time?

Name _____

Fill in the Blanks

Complete the sentences with one of the common time expressions at the bottom of the page.

1. Come and help me with the garbage, since you seem to have so much _____.

2. _____ there was a beautiful princess who lived in a castle in the clouds.

3. Why wait? _____.

4. I've told you, _____, not to bite your nails.

5. I can't believe it! We were _____ for a doctor's appointment.

6. Well, _____. I've been waiting here for almost an hour!

7. That camping trip last summer was terrific. I had _____.

8. It will get done, my dear, I assure you. _____.

9. Quit _____ and get on with your homework.

10. She made it around the track in _____.

a. time and time again
b. once upon a time
c. it's about time
d. record time
e. time on your hands
f. wasting time
g. there's no time like the present
h. all in good time
i. on time
j. the time of my life

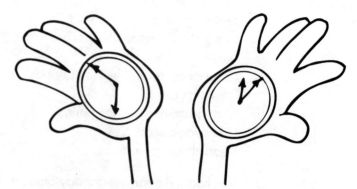

Timely Phrases

Mix 'n' Match

Draw a line to match each timely phrase with its meaning.

Time is up   beyond the reach of memory
Time out   for now
In the nick of time   occasionally
From time to time   just
Sometimes   now and then
For the time being   a pause
Time immemorial   over

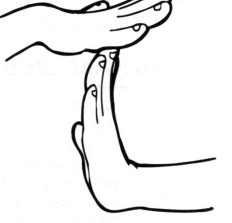

The "Write" Time

Pick five phrases from this list. Write a sentence using each phrase.

- Time flies when you're having fun.
- Time marches on.
- There is a time and a place for everything.
- Time is money.
- Time stands still.
- Passing the time.
- Killing time.
- Time is running out.

_____

_____

_____

_____

_____

_____

The Time Talk Test

Listed below are some timely expressions that you might have heard before. What do you think each one means?

1. A stitch in time saves nine. _____

2. A race against time. _____

3. The sands of time. _____

4. Time heals all wounds. _____

5. Time waits for no man. _____

Time Management

One of the most difficult things about time, is managing it properly–getting places on time, getting everything done that you want to do, spending your time wisely and efficiently. It always seems that there is so much to do and so little time. Getting organized will help you to get the most out of every second.

These days time management experts are making a very good living telling other people how to organize their time. Some people believe that the more organized you are, the more successful you are and the more time you will have left-over to do the things you really *want* to do.

"To do great important tasks, two things are necessary: a plan and not quite enough time."

A Day in the Life of . . .

We all live our lives according to the same 24-hour clock, seven-day week, 12-month calendar. But when it comes down to the nitty gritty, it is amazing how differently we spend our time! Discuss the "repetitive" things you do every school day. You know, things like: waking up, eating breakfast, leaving for school, arriving home, eating dinner, doing homework, watching TV, going to bed. Keep a "log" of your activities.

Things to Think About
- How does your weekday schedule differ from your weekend schedule?
- How does your weekday/weekend schedule differ from your classmates'? What things do you all do?
- Compared to your classmates, do you lead a busy life or a more relaxed one?
- What is the most interesting thing you do each week?
- Ask your parents to keep a log of their activities. Whose schedule is the most hectic? Whose is the most enjoyable?

By a show of hands, find out who in the group:
 watches the most TV
 spends the most time reading, playing sports
 spends the most time getting to/from school
 spends the most time doing homework
Think of other interesting "contests."

Fast Fact: Congratulations! If you have completed the *My Week Booklet* on page 131, or if you kept a diary of your own, you have written your own history book. These are permanent records of history that may seem boring and ordinary to you but would be most interesting to children in the future if they are found one day!

130

My Week Booklet

Keep a one-week diary using the boxes below. Draw a picture and write a few words or sentences in each box to describe what you did each day of the week. Cut out the boxes and staple together to make a booklet.

Staple here.

My Week

by _____

date: _____   1

Staple here.

Monday

2

Staple here.

Tuesday

3

Staple here.

Wednesday

4

Staple here.

Thursday

5

Staple here.

Friday

6

Staple here.

Saturday

7

Staple here.

Sunday

8

131

How Do You Spend Your Time?

How do you spend your time? Fill in the blanks below to find out!

Days

I spend:

____	hours each day at school.
____	hours each day sleeping.
____	hours each day eating.
____	hours each day watching TV.
____	hours each day at recess.
____	hours each day doing homework.
____	hours each day playing sports.
____	hours each day getting to/from school.
____	hours each day doing chores.
____	hours each day playing.
____	hours each day in the car.
____	hours each day playing with friends.
____	hours each day arguing and fighting.
____	hours each day sitting.
____	hours each day reading.

Weeks

____	hours each week.
____	hours each week.
____	hours each week.
____	hours each week.
____	hours each week.
____	hours each week.
____	hours each week.
____	hours each week.
____	hours each week.
____	hours each week.
____	hours each week.
____	hours each week.
____	hours each week.
____	hours each week.
____	hours each week.

What do you spend the most time doing? _____

What do you spend the least time doing? _____

What would you like to do more? _____

What would you like to do less? _____

What else do you do every day or week that is not on this list? How much time do you spend doing it?

_____

_____

At this rate, how much of your life will you spend sleeping? _____ Eating? _____

To figure this out, first multiply the weekly amount by 52 (because there are 52 weeks in a year). Then multiply this "annual" figure by 75 (because most people live to be about 75 years of age).

Fast Fact: By the time you are 75 years old, you will probably have walked 65,000 miles. That is the equivalent of four times around the Earth!

Time Capsules: Message in a Bottle

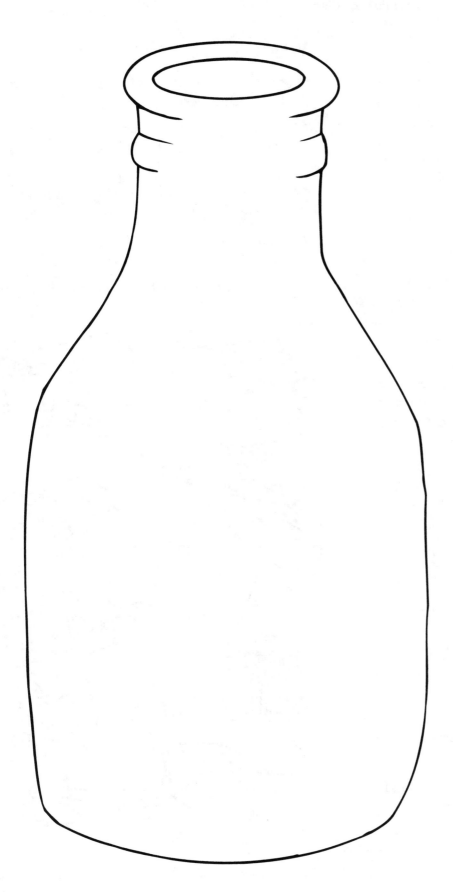

Wouldn't it be great to find a message in a bottle that was buried for hundreds, maybe even thousands, of years? Wouldn't it be even greater to bury your own message in a bottle and have it discovered hundreds or even thousands of years from now?

People often leave messages for future generations. Some are meant to be discovered; others are not. Historical documents are often kept in vaults for safekeeping and future reference. But many relics of the past are discovered by accident. Like the beautiful artifacts–and even bodies!–carefully concealed and preserved by the ancient Egyptians. Although the great pyramids the Egyptian kings were intended as sacred burial grounds, they are often regarded as giant time capsules that give our society clues about a civilization that existed thousands of years ago.

Time capsules are like small (or, in the case of the pyramids, huge!) museums. When they are discovered, it is as if a door has suddenly been opened onto the secret world of the past. The time capsules that you bury today will help future scientists and historians understand more about what life was like for kids in the 20th century.

Fast Fact: You probably know that robotic spacecraft take pictures of the planets in our solar system and send important information back to Earth. But did you also know that these same craft also act as earthly ambassadors to distant intergalactic civilizations? It's true. The space probe *Voyageur 2*, which visited Saturn, Uranus and Neptune before passing out of our solar system in 1990, carries a symbolic message from the people of Earth. It is hoped that this peaceful greeting will someday be discovered by curious and friendly interplanetary visitors.

133

Class Capsules

K-4 • History/Science/Language/Health • Creative Thinking/Analysis/Group Interaction/Prediction/Reading and Writing • 60 min +

Make your own group time capsule!

Materials

large glass jar (or other container that will not disintegrate over time)
"I Am Me" class collection of forms and personal items
secret burial or hiding place (On school property? Municipal property? Private property?)
shovel

Get Started

1 Think about what it is that makes you who you are. What does being a kid mean to you? What things would you show/tell to a visitor from outer space to help him/her/it understand who you–and other kids your age–are?

2 Have each child bring at least one small item that is special to them to be buried for reference by future generations! This will be a personal contribution to the class's "We Are We" collection.

- "We Are We" collection suggestions:

- a postcard of your tow

- a local map (hand-drawn or reproduced)

- photographs or hand-drawn pictures of you doing things you like to do, where you like to do them

- a coin or commemorative stamp

- collector's cards

- the front page of your local newspaper

- a magazine or newspaper article that describes life as it is for kids today's

- a recent poll or survey (professional or amateur) of kids' favorite foods, movies, television shows

- a toy or trinket

3 Find a place to bury the time capsule. Be sure to get permission before you dig!

What to Do

1 Have each participant fill in "The I Am Me Personal History Document."

2 Write a cover letter with the group explaining what you are doing and why. Have one student or yourself act as the scribe (writer).

3 Place all of the "I Am Me" information sheets and personal items in the jar. Screw the lid on tightly.

4 Have a ceremonial dig to bury the time capsule.

5 Sit back and wait for the future to unfold!

The "I Am Me" Personal History Document

My name is _____

I am _____ years of age.

I was born in the city/town of _____

I live with _____

The name of my school is: _____

I am in grade: _____

I spend most of my time _____

I like to _____

I don't like to _____

I like to eat _____

I like to drink _____

I go to bed at _____

I like to play _____

The sports/hobbies I like best are _____

My favorite television shows are _____

My favorite movies are _____

My favorite books are _____

I am afraid of _____

The worst problems facing the world today are: _____

_____

I would like to change the world by _____

_____

When I grow up I would like to _____

_____

The best thing about being a kid today is _____

_____

Time Card Bingo

Materials

at least two players and one caller
Time Bingo Cards
Time Bingo Master Clock Sheet
bingo chips (pennies, pebbles or beans)

Get Started

1 Each player selects a Time Card and takes a handful of bingo chips.

2 Each player puts a chip on the FREE space in the middle of the card.

3 Choose a caller to call and cross off "time facts" on the Master Clock Sheet below.

How to Play

1 Caller chooses a fact calls it out and crosses it off of the Master Clock Sheet.

2 When a time fact matches a time space on the player's Time Card, player marks the spot with a bingo chip.

3 Play continues in this way until one player calls out "Bingo!"

4 Caller checks the Master Clock Sheet to see that all the times marked on the player's card have been called.

5 Play continues with remaining players or a new game begins.

Time Card Bingo Master Clock Sheet

Cross out the items as they are called. (Answers are in parentheses. Do not read this information aloud.)

1:00	4:00	7:30	10:30	1:15	4:15	7:45	10:45
2:00	5:00	8:30	11:30	2:15	5:15	8:45	11:45
3:00	6:00	9:30	12:30	3:15	6:15	9:45	12:45

morning hours (a.m.)

after noon hours (p.m.)

60 seconds in a (minute)

60 minutes in a (hour)

$1/60$ of an hour (minute)

24 hours in a (day)

a wrist clock (watch)

time when the sun is directly overhead (noon)

system for organizing the days and years (calendar)

a kind of shadow clock (sundial)

10 years (decade)

100 years (century)

1,000 years (millennium)

the day before today (yesterday)

the day after today (tomorrow)

365 $1/4$ days (year)

the most accurate clock (atomic clock)

leap year occurs every (4 years)

days in a lunar month (29 $1/2$)

days in a leap year (366)

months in a year (12)

the first month (January)

the last month (December)

the sixth month (June)

our calendar (Gregorian)

12 hours of daylight and darkness (equinox)

smallest whole unit of time measurement (second)

rotation of the Earth takes (24 hours)

hourglass

sun

moon

cuckoo clock

pendulum

136

Time Card Bingo

1:00	a.m.	hour	
24 hours	sundial	9:30	century
June	366		yesterday
watch	atomic clock	8:45	equinox

Time Card Bingo

p.m.	12:45	year	December
4:00		cuckoo clock	January
Gregorian	5:00	6:15	millennium
noon	decade		minute

Time Card Bingo

	7:30	tomorrow	second
equinox	calendar	century	4 years
11:45		a.m.	January
day	29 1/2	3:15	

Time Card Bingo

second	12	atomic clock	5:15
	yesterday		watch
minute	7:45	sundial	December
hour	p.m.	366	6:00

Time Card Bingo

3:00	4 years	noon	
Gregorian	decade	$29\frac{1}{2}$	June
24 hours	12	FREE	millennium
day	2:00	9:45	minute

Time Card Bingo

8:30	12:30	equinox	century
2:15	FREE	cuckoo clock	calendar
second	1:15	4:15	Gregorian
decade	11:30		atomic clock

Time Card Bingo

	2:15	calendar	noon
January	24 hours	minute	10:45
10:30	FREE	year	tomorrow
8:30	366	5:15	

Time Card Bingo

watch	4 years	December	1:15
	June	FREE	yesterday
$29\frac{1}{2}$	8:45	sundial	minute
hour	p.m.	10:30	5:00

138

Time Travelers Board Game

Materials

a die
enlarged photocopied gameboard
1 game piece per player
Time Out! cards

How to Play

1 Players put their game pieces on Start.

2 Time Out! cards are shuffled and placed, face-down on the Time Out! place on board.

3 Youngest player rolls die. Player matches die to Clock Box and moves game piece to closest time. (For example: If the player's game piece is on 2:15 at start of turn and a two is rolled, player must move to 2:35. If game piece is on 2:45 at start of turn, player would move to 3:05.)

4 If player lands on a dark space, player draws a Time Out! card and follows instructions.

5 Play continues, clockwise, until one player reaches 12:00 with an exact roll.

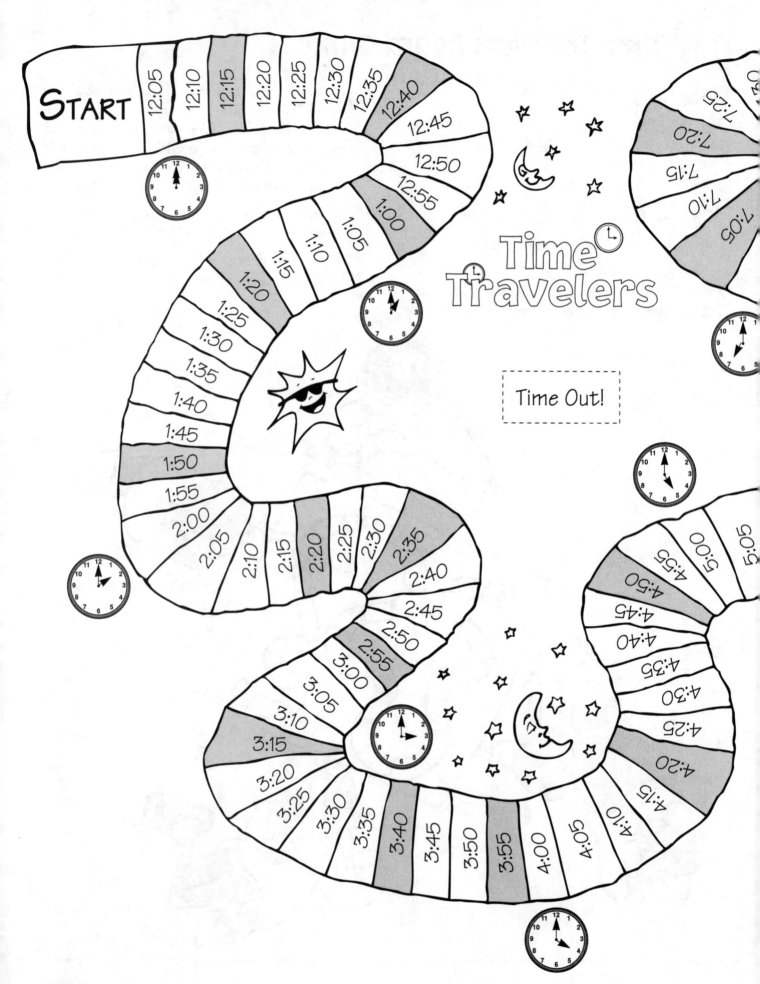

START

12:05 12:10 12:15 12:20 12:25 12:30 12:35 12:40 12:45 12:50 12:55 1:00 1:05 1:10 1:15 1:20 1:25 1:30 1:35 1:40 1:45 1:50 1:55 2:00 2:05 2:10 2:15 2:20 2:25 2:30 2:35 2:40 2:45 2:50 2:55 3:00 3:05 3:10 3:15 3:20 3:25 3:30 3:35 3:40 3:45 3:50 3:55 4:00 4:05 4:10 4:15 4:20 4:25 4:30 4:35 4:40 4:45 4:50 4:55 5:00 5:05

7:05 7:10 7:15 7:20 7:25

Time Travelers

Time Out!

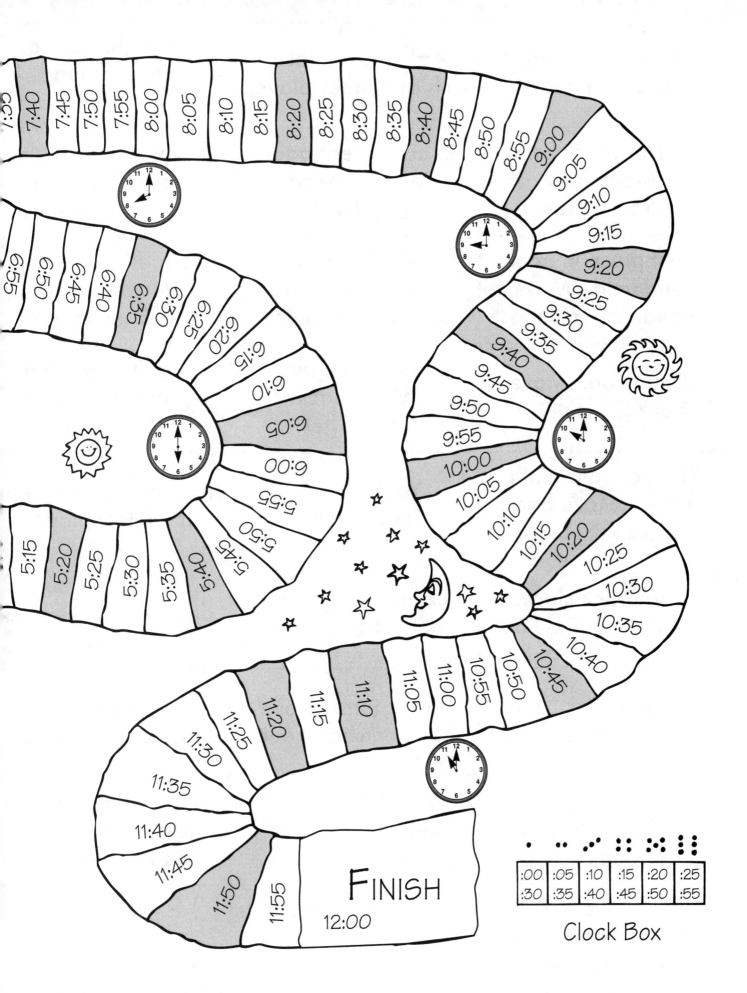

FINISH

12:00

Clock Box

:00	:05	:10	:15	:20	:25
:30	:35	:40	:45	:50	:55

141

Late for school. Miss one turn.	A+! Trade places with any other player.	Daylight Saving. Move ahead one hour.
Clean your room. Move back 55 minutes.	Read a great book. Move ahead 55 minutes.	Science project due tomorrow. Move back 50 minutes.
First place in track and field. Move ahead 50 minutes.	Violin lesson canceled. Move ahead 45 minutes.	School's out early. Move ahead 35 minutes.
Detention. Move back 20 minutes.	Missed the bus. Move back 15 minutes.	Don't forget the garbage. Move back 10 minutes.
Forgot your math book. Move back five minutes.	You're early. Roll again.	F! Trade places with the latest player.
Turn back clock. Move back one hour.	Grounded. Move back 45 minutes.	Brussel sprouts disaster. Move back 40 minutes.
No homework. Move ahead 30 minutes.	Traffic jam. Move back one-half hour.	Snowstorm. Move back 35 minutes.
You chose the quick line! Move ahead one-half hour.	Time off for good behavior. Move ahead 20 minutes.	Finished your homework. Move ahead 15 minutes.
Ran all the way to school. Move ahead 10 minutes.	Took a shortcut. Move ahead five minutes.	

TLC10073 Copyright © Teaching & Learning Company, Carthage, IL 62321-0010

Answer Key

Page 35

Threatened
1. Spiny Soft-Shell Turtle
2. Black Red Horse
3. Burrowing Owl
4. Pine Marten

Endangered
1. Blue Racer
2. Aurora Trout
3. Whooping Crane
4. Beluga Whale

Extirpated
1. Pygmy Short-horned Lizard
2. Paddlefish
3. Greater Prairie Chicken
4. Swift Fox

Extinct
1. Blue Walleye
2. Passenger Pigeon
3. Sea Mink
4. Great Auk

Page 49
1. 360°, 24, north and south
2. Zero or prime meridian
3. 1:00 p.m.
4. 11:00 a.m.
5. 12:00 midnight, Monday
6. 12:00 midnight, Sunday
7. International Date Line, each day begins and ends at midnight on this line.

Page 55

1. F		12. H	
2. O		13. T	
3. B		14. K	
4. E		15. P	
5. D		16. C	
6. M		17. U	
7. S		18. L	
8. R		19. N	
9. I		20. A	
10. G		21. Q	
11. J			

Page 78
1. 2:00
2. 6:00
3. 8:00
4. 4:00
5. 10:00
6. 3:00
7. 9:00
8. 2:00

Page 80

Half Past
1. 6:30
2. 2:30
3. 12:30

Quarter After
1. 1:15
2. 8:15
3. 9:15

Quarter To
1. 1:45
2. 10:45
3. 5:45

Page 81

Before 12:30
1. 1:30
2. 5:15
3. 2:10
4. 4:15

After 12:30
1. 3:05
2. 3:45
3. 5:50
4. 4:35

Page 82
1. 12, 60
2. Minute
3. Hour
4. 5
5. 60
6. 24
7. 60
8. 30
9. 15

Page 84
1. one fifteen or quarter after one
2. eleven thirty
3. two forty-five or quarter to three
4. three twenty or twenty after three
5. 1:00
6. 12:00
7. 3:30
8. 10:15
9. 6:30-half past six
10. 6:00-six o'clock
11. 12:00-twelve noon
12. 3:00-three thirty

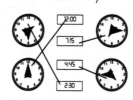

Page 85

What Time Is It?
1. 12:00
2. ten thirty or half past ten
3. 3:30
4. one ten or ten past one

The Time Is . . .
5.
6. 3:10
7.
8. 12:00
9. 6:00
10. 9:30
11. 3:45
12. 5:55
13. 7:00
14. 5:00

Page 86
Twelve o'clock or noon is left.

Page 87
1. 12
2. 60
3. 30
4. 15
5. 6:00
6. 10:30
7. 9:15
8. 1:30
9.  10.
11. 12.
13. 6:30
14. 3:00
15. 4:55
16. 11:10

Page 128
1. E
2. B
3. G
4. A
5. I
6. C
7. J
8. H
9. F
10. D

TLC10073 Copyright © Teaching & Learning Company, Carthage, IL 62321-0010

143

Bibliography

Boslough, John. "Enigma of Time," *National Geographic*, March 1990, pp. 109-32.

Darling, Dr. David. *Could You Ever Build a Time Machine?* Dillon Press, Inc. Minneapolis, Minnesota 55415, 1991.

Galt, Tom. *Seven Days from Sunday*. Thomas Y. Crowell Company, Binghamton, New York, 1956.

Gega, Peter C. *Science in Elementary Education*, 6th edition, Macmillan Publishing Company, New York, 1990.

Griffin, John and Mary. *Time and Space*. Stoddart Publishing Co. Limited, Toronto, Canada, 1994.

Hawking, Stephen, W. *A Brief History of Time; From the Big Bang to Black Holes*, Bantam Books, Toronto/New York/London/Sydney/Auckland, 1988.

Kramer, Ann (senior editor). *The Random House Children's Encyclopedia*. Dorling Kindersley Limited, London, England, 1991.

Magdalen Bear, Lou Pamenter, *Canadian Days; A Calendar for All Time*. Pembroke Publishers Limited, Ontario, Canada, 1990.

Morris, Johnny. *The Animal Roundabout*. Stoddart Publishing Co. Limited, Toronto, Canada, 1993.

Neal, Harry Edward. *The Mystery of Time*. Julian Messner, New York, 1966.

Norris, Doreen, and Joyce Boucher. *Observing Children in the Formative Years*. The Board of Education for the City of Toronto, Toronto, 1989.

Parry, Caroline. *Let's Celebrate! Canada's Special Days*. Kids Can Press Ltd., Toronto, 1987.

Riedman, Sarah H. *Biological Clocks*. Thomas Y. Crowell Junior Books, New York/Fitzhenry and Whiteside Limited, Toronto, 1982.

Science Framework for California Public Schools, Kindergarten Through Grade Twelve, Adopted by the California State Board of Education, California Department of Education, Sacramento, 1990.

Siegel, Alice, and Maro McLoone Basta, *The Information Please Kids' Almanac*. Houghton Mifflin Company. Boston/New York/London, 1992.

The Guinness Book of Records, Bantam Books, New York, New York, 1996.